FINANCE POLICY *for* RENEWABLE ENERGY *and a* SUSTAINABLE ENVIRONMENT

ENERGY AND THE ENVIRONMENT

SERIES EDITOR

Abbas Ghassemi

New Mexico State University

PUBLISHED TITLES

Finance Policy for Renewable Energy and a Sustainable Environment
Michael Curley

Wind Energy: Renewable Energy and the Environment, Second Edition
Vaughn Nelson

Solar Radiation: Practical Modeling for Renewable Energy Applications
Daryl R. Myers

Solar and Infrared Radiation Measurements
Frank Vignola, Joseph Michalsky, and Thomas Stoffel

Forest-Based Biomass Energy: Concepts and Applications
Frank Spellman

Introduction to Renewable Energy
Vaughn Nelson

Geothermal Energy: Renewable Energy and the Environment
William E. Glassley

Solar Energy: Renewable Energy and the Environment
Robert Foster, Majid Ghassemi, Alma Cota
Jeanette Moore, and Vaughn Nelson

FINANCE POLICY *for* RENEWABLE ENERGY *and a* SUSTAINABLE ENVIRONMENT

Michael Curley

CRC Press
Taylor & Francis Group
Boca Raton London New York

CRC Press is an imprint of the
Taylor & Francis Group, an **informa** business

CRC Press
Taylor & Francis Group
6000 Broken Sound Parkway NW, Suite 300
Boca Raton, FL 33487-2742

Printed on acid-free paper
Version Date: 20140127

International Standard Book Number-13: 978-1-4398-9419-4 (Hardback)

Library of Congress Cataloging-in-Publication Data

Curley, Michael.
Finance policy for renewable energy and a sustainable environment / Michael Curley.
pages cm
Summary: "This book explains how environmental projects and improvements are achieved through the imposition of regulations, on the one hand, and financial incentives on the other. It discusses how those incentives can be organized to achieve the greatest environmental benefits at the lowest possible cost to the public. The book presents the best environmental finance policies to bear on the financing of alternative energy projects so that the ultimate cost of delivered power will decline. It also examines the challenges of the next generation of environmental programs"-- Provided by publisher.
Includes bibliographical references and index.
ISBN 978-1-4398-9419-4 (hardback)

1. Renewable natural resources. 2. Sustainable development--Environmental aspects. 3. Sustainable development--Finance. I. Title.
HC85.C87 2014 333.79'4--dc23

Visit the Taylor & Francis Web site at
http://www.taylorandfrancis.com

and the CRC Press Web site at
http://www.crcpress.com

This book is dedicated to George F. Ames, who runs the Clean Water State Revolving Fund (CWSRF) program at the US Environmental Protection Agency (EPA). The CWSRF is 25 years old and has provided over $100 billion of financial assistance for clean water projects. It is the most successful environmental finance program on planet Earth.

George not only runs the CWSRF, but also was part of a team of three who thought it up and drafted Title VI of the Clean Water Act that authorizes the program.

George has held several other major positions at the EPA; but, in his spare time there, he also helped invent the Environmental Financial Advisory Board, created Environmental Finance Centers at nine major universities and took a leave of absence to create the Council of Infrastructure Financing Agencies (CIFA), which is now the association of all the individual state CWSRFs.

Contents

Series Preface

By 2050 the demand for energy could double or even triple as the global population rises and developing countries expand their economies. According to data from the United Nations, it is projected that the world population will increase from 7.2 billion to more than 9 billion by 2050. This increase, coupled with continued demand for the same limited natural resources, will cause a significant increase in consumption of energy. All life on Earth depends on energy and the cycling of carbon. Affordable energy resources are essential for economic and social development as well as food production, water supply availability, and sustainable healthy living. In order to avoid long-term adverse and potentially irreversible impacts of harvesting energy resources, we must explore all aspects of energy production and consumption, including energy efficiency, clean energy, global carbon cycle, carbon sources and sinks, and biomass, as well as their relationship to climate and natural resource issues. Knowledge of energy has allowed humans to flourish in numbers unimaginable to our ancestors. The world's dependence on fossil fuels began approximately 200 years ago. Are we running out of oil? No, but we are certainly running out of the affordable oil that has powered the world economy since the 1950s. We know how to recover fossil fuels and harvest their energy for operating power plants, airplanes, trains, and automobiles, which results in modifying the carbon cycle and additional greenhouse gas emissions. This has resulted in debate on the availability of fossil energy resources, peak oil era and timing for an anticipated end of the fossil fuel era, and price and environmental impact versus various renewable resources and use, carbon footprint, emission, and control, including cap-and-trade and the emergence of "green power."

Our current consumption has largely relied on oil for mobile applications, and coal, natural gas, nuclear, or water power for stationary applications. In order to address the energy issues in a comprehensive manner, it is vital to consider the complexity of energy. Any energy resource, including oil, gas, coal, wind, biomass, etc., is an element of a complex supply chain and must be considered in the entirety as a system from production through consumption. All the elements of the system are interrelated and interdependent. Oil, for example, requires consideration for interlinking all the elements, including exploration, drilling, production, transportation, water usage and production, refining, refinery products and by-products, waste, environmental impact, distribution, consumption/application, and finally emissions. Inefficiency in any part of the system has an impact on the overall system, and disruption in one of these elements causes a major interruption and a significant cost impact. As we have experienced in the past, interrupted exploration will result in disruption in production, restricted refining and distribution, and consumption shortages; therefore, any proposed energy solution requires careful evaluation and, as such, may be one of the key barriers to implementing the proposed use of hydrogen as a mobile fuel.

Although an admirable level of effort has gone into improving the efficiency of fuel sources for the delivery and use of energy, we are faced with severe challenges on many fronts. These include population growth, emerging economies, new and expanded usage, and limited natural resources. All energy solutions include some level of risk, including technology snafus, changes in market demand, economic drivers, and others. This is particularly true when proposing energy solutions involving the implementation of untested alternative energy technologies.

There are concerns that emissions from fossil fuels lead to climate changes with possibly disastrous consequences. Over the past five decades, the world's collective greenhouse gas emissions have increased significantly, even as efficiency has increased, resulting in extending energy benefits to more of the population. Many propose that we improve the efficiency of energy use and conserve resources to lessen greenhouse gas emissions and avoid a climate catastrophe. Using fossil fuels more efficiently has not reduced overall greenhouse gas emissions for various reasons, and it is unlikely that such initiatives will have a perceptible effect on atmospheric greenhouse gas content. While there is a debatable correlation between energy use and greenhouse gas emissions, there are effective means to produce energy, even from fossil fuels, while controlling emissions. There are also emerging technologies and engineered alternatives that will actually manage the makeup of the atmosphere, but will require significant understanding and careful use of the energy.

We need to step back and reconsider our role and knowledge of energy use. The traditional approach of micromanagement of greenhouse gas emissions is neither feasible nor functional over a long period of time. More assertive methods to influence the carbon cycle are needed and will be emerging in the coming years. Modifications to the carbon cycle mean that we must look at all the options in managing atmospheric greenhouse gases, including various ways to produce, consume, and deal with energy. We need to be willing to face reality and search in earnest for alternative energy solutions. There appear to be technologies that could assist; however, they may not all be viable. The proposed solutions must not be in terms of a "quick approach," but rather a more comprehensive, long-term (10, 25, and 50+ years) approach that is science based and utilizes aggressive research and development. The proposed solutions must be capable of being retrofitted into our existing energy chain. In the meantime, we must continually seek to increase the efficiency of converting energy into heat and power.

One of the best ways to define sustainable development is through long-term, affordable availability of limited resources, including energy. There are many potential constraints to sustainable development. Foremost among these is the competition for water use in energy production, manufacturing, farming, and others versus a shortage of fresh water for consumption and development. Sustainable development is also dependent on the Earth's limited amount of productive soil. In the not-too-distant future, it is anticipated that we will have to restore and build soil as part of sustainable development. We need to focus our discussions on the motives, economics, and benefits of natural resource conservation, as well as the limitation of technology improvement in impacting sustainability; that is, we are limited in catching fish from the ocean due to the number of fish that are available, not bigger boats or better nets. Hence, possible sustainable solutions must not be solely based on technology enhancement and improvement, specifically in obtaining the fossil resources, but be comprehensive and based on integrating our energy use with nature's management of carbon, water, and life on Earth as represented by the carbon and hydrogeological cycles. The challenges presented by the need to control atmospheric greenhouse gases are enormous and require "out-of-the-box" thinking, innovative approaches, imagination, and bold engineering initiatives in order to achieve sustainable development. We will need to ingeniously exploit even more energy and integrate its use with control of atmospheric greenhouse gases.

The continued development and application of energy are essential to the sustainable advancement of society. Therefore, we must consider all aspects of the energy options, including performance against known criteria, basic economics and benefits, efficiency, processing and utilization requirements, infrastructure requirements, subsidies and credits, waste and ecosystem, as well as unintended consequences such as impacts on natural resources and the environment. Additionally, we must include the overall changes and the emerging energy picture based on current and future efforts in renewable alternatives, modified and enhanced fossil fuels, and evaluate the energy return for the investment of funds and other natural resources such as water. Water is a precious commodity in the West in general and the Southwest in particular, and has a significant impact on energy production, including alternative sources due to the nexus between energy and water and the major correlation with the environment and sustainability-related issues.

A significant driver in creating this book series that is focused on alternative energy and the environment was a consequence of lecturing around the country and in the classroom on the subject of energy, the environment, and natural resources such as water. While the correlation between these elements, how they relate to each other, and the impact of one on the other is understood, it is not significantly debated when it comes to the integration and utilization of alternative energy resources into the energy matrix. Additionally, as renewable technology implementation grows in various states, nationally, and internationally, the need for informed and trained human resources continues to be a significant driver in future employment, resulting in universities, community colleges, and trade schools offering minors, certificate programs, and even in some cases majors in renewable energy and sustainability. As the field grows, the demand for trained operators, engineers, designers, and architects who would be able to incorporate these technologies into their daily activity is increasing. We receive a daily deluge of flyers, e-mails, and texts on various short courses available for interested parties in solar, wind, geothermal, biomass, etc., under the umbrella of re-tooling an individual's career and providing trained resources needed to interact with financial, governmental, and industrial organizations.

In all my interactions throughout the years in this field, I have conducted significant searches to locate integrated textbooks that explain alternative energy resources in a suitable manner and that would complement a syllabus for potential courses to be taught at the university while providing good reference material for interested parties getting involved in this field. I have been able to locate a number of books on the subject matter related to energy, energy systems, resources such as fossil nuclear, renewable and energy conversion, as well as specific books on the subjects of natural resource availability, use, and impact as related to energy and environment. However, specific books that are correlated and present the various subjects in detail are few and far between. We have therefore started a series of books, each addressing specific technology fields in the renewable energy arena. As a part of this series, there are textbooks on wind, solar, geothermal, biomass, hydro, and others yet to be developed. Our books are intended for upper-level undergraduate and graduate students, and for informed readers who have a solid fundamental understanding of science and mathematics as well as individuals/organizations that are involved with design development of renewable energy field entities and are interested in having reference material available to their scientists and engineers, consulting organizations, and reference libraries. Each book presents fundamentals as well as a series of numerical and conceptual problems designed to stimulate creative thinking and problem solving.

The series editor wishes to express his deep gratitude to his wife Maryam, who has served as a motivator and intellectual companion and too often was victim of this effort. Her support, encouragement, patience, and involvement have been essential to the completion of this series.

Abbas Ghassemi, PhD
Las Cruces, New Mexico

About the Series Editor

Abbas Ghassemi, PhD, is the director of the Institute for Energy and the Environment (IEE) and professor of chemical engineering at New Mexico State University. As the director of IEE, he is the chief operating officer for programs in education and research, and outreach in energy resources, including renewable energy, water quality and quantity, and environmental issues. He is responsible for the budget and operation of the program. Dr. Ghassemi has authored and edited several textbooks and has written many publications and papers in the areas of energy, water, carbon cycle including carbon generation and management, process control, thermodynamics, transport phenomena, education management, and innovative teaching methods. His research areas of interest include risk-based decision making, renewable energy and water, carbon management and sequestration, energy efficiency and pollution prevention, multiphase flow, and process control. Dr. Ghassemi serves on a number of public and private boards, editorial boards, and peer review panels and earned MS and PhD degrees in chemical engineering, with minors in statistics and mathematics, from New Mexico State University and a BS degree in chemical engineering, with a minor in mathematics, from the University of Oklahoma.

Preface

This is not your average textbook. It is not a compilation of dry facts and drier theories. I started looking into environmental finance issues about 25 years ago. In many cases, I was appalled at what I saw. This book is a reflection of those experiences.

I come from Wall Street. I have seen how business deals with financial issues. Financial markets are highly efficient. Government has much to learn.

Before Wall Street, I was in politics in New York State. Winston Churchill said, "Democracy is the worst form of government, except for all others that have been tried." He also said, "The best argument against democracy is a five-minute conversation with the average voter." I am not so cynical, but I do see clearly that the exigencies of politics temper all good-hearted attempts to bring business-like efficiencies to environmental finance. I understand why a state legislator would vote for a grant program, even when it is wasteful and foolhardy to do so. I sympathize with them; but that is not going to stop me from pointing out the folly of their actions.

When colleagues asked me the working title of the book I was writing, I, of course, told them, "*Finance Policy for Renewable Energy and a Sustainable Environment*." Their next question invariably was, "When did you start writing fiction?"

Their point is well taken. There is no finance policy for renewable energy. And, there is no finance policy for a sustainable environment.

When I teach environmental finance, I often facetiously tell my class that they are listening to the world's greatest expert in environmental finance... under one theory that, in the land of the blind, the one-eyed man is king!

This point is also well taken. I have been practicing what I call "environmental finance" for the past 25 years. During that time, I have never met another living soul who said that he, or she, was practicing environmental finance. There are over 17,000 employees at the United States Environmental Protection Agency (EPA). After working with that agency for 25 years, I would guess that fewer than one-tenth of 1% of them—about 17—know what environmental finance is. Don't get me wrong. Everyone whom I have met at the EPA is a fine and decent person. They are unquestionably devoted to improving the environment. It is just that the vast majority of people—not just at the EPA, but in the whole country—regard the EPA as a regulatory agency. It certainly is that; but it is also home to the largest and most successful environmental finance program in the world. The EPA people know a lot about sticks, but little about the carrots that need to accompany the sticks.

"Environmental finance is about creating the greatest environmental benefit for the largest number of people at the lowest possible cost." I put this sentence in quotes because it is the most important sentence in this book. In

Chapter 2, I begin laying out what I think are 23 principles of environmental finance. You will see that the sentence above, in quotes, is the first core principle and, by far, the most important. You will also see it repeated often in this book.

And so, this book is about providing the greatest environmental benefit for the largest number of people at the lowest possible cost.

Achieving environmental quality in an evermore-populous world will be expensive. So too, will be the journey from fossil fuels to renewables. But we must do both. And we must do so in the most painless way possible.

Finance Policy for Renewable Energy and a Sustainable Environment is about creating an effective environmental finance policy. It is about raising funds in the least painful way possible, changing behavior for the good as we do so. It is about spending money wisely to achieve a better quality of life. It is about making financial decisions based on science, not politics. And, it is about designing and executing environmental finance programs in the most cost-effective manner possible. It is about making policy that will gain us the greatest environmental benefits for the largest number of people at the lowest possible cost. It is about the best for the most for the least.

I say that there is no "finance policy for renewable energy and a sustainable environment." This is true at the global level, the national level in the United States, and at the level of most states as well. This is not surprising. There is no energy policy at all in the United States at this time, much less a renewable energy policy. Some want cheap energy and energy security/independence. Others would like to see more renewable energy and clean energy. But they are very expensive. Because they are expensive, they must be subsidized to evade criticism and be broadly acceptable. But the money for the subsidies has to come from somewhere. We have to "rob Peter to pay Paul," but no one wants to get caught. No one wants to raise any taxes or fees or create new ones. Furthermore, many people are intimidated by claims about energy security, which are true. If we have to rely on renewables alone, we will not achieve energy security even in the remote future, much less any foreseeable future.

So, this book will attempt to point out what exists today in terms of how environmental quality is being paid for. And, more importantly, it will point the way to what needs to be done to assure a quality environment for the generations to come.

Finally I need to say something about the state of Maryland. I moved from New York City to Maryland in 1992. At that time, I was doing research aimed at founding a financial guaranty insurance company to insure the long-term debt of water and wastewater systems. As fate would have it, the Clean Water Act was amended in 1987 to replace the $70+ billion construction (of wastewater treatment plants) grants program with a loan program called the Clean Water State Revolving Fund. The EPA, charged with managing the new program, was in a quandary. The individual states were supposed to make loans to wastewater systems but there was no one in any state government who had ever done so before. So, the EPA people found me lurking around asking questions about default rates, learned I was a finance guy, and recruited me to teach the state people how to make loans and the local people how to apply for them and administer them. All those materials eventually became a book, *The Handbook of Project Finance for Water and Wastewater Systems*, which I wrote for Lewis Publishers, who were eventually acquired by Taylor & Francis, the publishers of this book. As a reward, the EPA folks got me appointed to their Environmental Financial Advisory Board (EFAB). When I moved to Maryland, the EFAB staff said that they needed an academic resource nearby. So, they asked me to set up an Environmental Finance Center. With the help of then-Governor William Donald Schaefer, we established the Center at the University of Maryland in College Park, right outside of Washington, DC.

From then on, I was plunged into the dynamic environmental community of Maryland. So, you will see many examples in this book of things that are happening (or not) in Maryland. One of the greatest concerns of the entire state is the health of the Chesapeake Bay. So, Maryland is very much into, and on the cutting edge of, environmental protection and conservation. As far as many states are concerned, and even many countries, Maryland is way ahead of the curve.

Good reading!

Acknowledgments

I need to mention some of the people to whom I go when I need advice and help, as I did many times during the writing of this book. Each has always been unfailingly gracious and generous with advice and assistance.

First and foremost is my good friend of 20+ years, Dr. John J. Boland, who is professor emeritus at the Johns Hopkins University. John is both a PE and a PhD in environmental economics. He has taught at Hopkins for more than 40 years, written more than 150 articles, and co-authored 4 books. He has consulted on environmental infrastructure projects all over the world. He served for six years on the US Environmental Protection Agency (EPA) Environmental Financial Advisory Board. John knows everything I want to know.

Next, in no particular order, is Jag Khuman, who is the director of the Maryland Water Quality Financing Administration, which runs Maryland's highly successful Clean Water State Revolving Fund and Safe Drinking Water Revolving Fund. Jag is an engineer and earned both an MBA and an MS in engineering management from Louisiana University. Jag knows everything there is to know about environmental infrastructure and how to pay for it.

Paul K. Marchetti has been the executive director of the Pennsylvania Infrastructure Authority (PennVest) since 1988. PennVest does all the environmental infrastructure financing in the Commonwealth. Paul has a PhD in economics from the University of Maryland and was president of the Council of Infrastructure Financing Agencies (CIFA) from 1994 through 1996. The CIFA is the national association for the 102 clean water and drinking water state revolving funds. Paul is an entrepreneur in government. Many states have wrestled with the critical issue of nutrient credit trading. Most have not gotten further than some limited point-source trades. Paul has developed the first full-fledged nutrient credit trading program in the country, covering both point-source and non-point-source credits.

I came late to air quality issues and even later to climate change. Robert K. Brenner is an expert on all of this. He taught me much more than I know. Among the many important duties he had, probably the most significant was director of the EPA's Office of Policy Analysis and Review from 1988 to 2011, where he played a key role in the enactment of the Clean Air Act Amendments of 1990, the most far-reaching environmental statute in the past 25 years. Rob received awards from President George H. W. Bush in 1993, President Bill Clinton in 1998, and President George W. Bush in 2003. He has bachelor's and master's degrees in economics and public policy from Princeton University's Woodrow Wilson School.

Last, but certainly not least, is Doug Siglin, who knows the human face of the struggle for environmental quality as well as the political one. He knows the impact of environmental finance on the federal budget as well as the family budget. Doug started out as a Peace Corps volunteer building clean water systems in villages in Zaire and went on to serve on the governing board of CARE. He started out as a congressional staffer and went on to direct the Washington efforts of four major environmental organizations: the Nature Conservancy; the World Wildlife Fund; American Rivers; and, from 2001 to 2013, he was the Chesapeake Bay Foundation's Washington representative and director of their Washington, DC office.

About the Author

Michael Curley is an attorney by training but has spent the majority of his career in finance, and the past 25 years of his career in environmental finance. He is currently a visiting scholar at the Environmental Law Institute in Washington, DC. During the 2012–2013 academic year, he was a visiting professor at the George Washington University School of Business, teaching courses in energy, environmental law, and environmental finance.

Mr. Curley was the founder and executive director of the International Center for Environmental Finance, which was funded with a $3 million grant from the US Environmental Protection Agency (EPA) and worked throughout the past decade in the former Soviet Union, Central America, and Asia to develop financial systems for environmental infrastructure and climate change projects. He has advised many governmental agencies and NGOs (non-governmental organizations); among them are the World Bank, the North Atlantic Treaty Organization (NATO), the Asian Development Bank, the Global Environment Facility, the US Agency for International Development, the United Nations Environment Program, the World Resources Institute, the National League of Cities, and the Shell (UK) Oil Foundation.

Mr. Curley is a pioneer in the financial guaranty insurance industry. In the 1980s, he raised the venture capital for and founded the third financial guaranty insurance company in the world, where he served as president and CEO until he sold the company to the Hong Kong Shanghai Bank.

Mr. Curley served as the president and CEO of the New York Job Development Authority, the state's economic development bank. He served as deputy commissioner and general counsel of the New York State Department of Economic Development. He was also counsel to the New York State Science and Technology Foundation, the state's venture capital agency. Prior to that, he served as the Parliamentarian of the New York State Assembly and associate counsel to the Speaker. Early in his career, he served as an assistant to Congressman Richard D. McCarthy in Washington, DC. After leaving government in 1979, he became a partner in the New York City law firm of Shea & Gould.

Mr. Curley is the author of a textbook, the *Handbook of Project Finance for Water and Wastewater Systems* (1993), published by Lewis Publishers. He has also written a monograph on environmental finance and more than

30 published articles. He founded the Environmental Finance Centers at the University of Maryland, Cleveland State University, and the Maxwell School at Syracuse University. He served as a Senior Lecturer at the Johns Hopkins University where he taught courses on finance and law in the energy and environmental fields, and he was also an adjunct professor of banking and finance at New York University where he taught venture capital as well as capital markets and investment banking. He has also taught environmental law and finance at the Vermont Law School. In 1990, he was appointed to the Environmental Financial Advisory Board (EFAB) at the EPA, where he served for 21 years under 4 US Presidents. He also served on the board of directors of the International Rural Water Foundation, and was co-chair of the Environmental Finance and Budget Subcommittee of Maryland Governor Martin O'Malley's Transition Team. He is on the advisory board of the Maryland–Asia Environmental Partnership and was a member of the Global Missions Committee of the Episcopal Diocese of Maryland. Governor William Donald Schafer appointed him to the Blue Ribbon Panel on the Financing of the Chesapeake Tributaries, and, while living in New York City, Mayor Edward Koch appointed him to the board of directors of the United Nations Development Corporation.

Introduction

This book is being written at a critical juncture in mankind's effort to come to grips with the environment. The first major environmental statute, the Clean Air Act (CAA), was passed in 1970. By 1980, the last of the "big five" environmental statutes, the Comprehensive Environmental Response, Compensation, and Liability Act (CERCLA), otherwise known as *the superfund law,* had been enacted. Between then and now, we have made major strides in improving environmental quality in the United States. Municipal sewage used to be the greatest nemesis of our nation's waterways. A generous construction grant program at the federal level enabled cities to build effective treatment works. Now, major water bodies, such as the Great Lakes, are clean enough to drink from without chlorine or boiling. Our air used to be burdened with massive quantities of lead. Through an aggressive vehicle emissions control program—and most importantly, by banning lead in gasoline—the ambient level of this toxin has been reduced by 94%. But, as it says in Chapter 1, these are the low-hanging fruit. And we have picked them all.

The next generation of pollution problems involves literally hundreds of thousands of small sources. What is more, is that these sources are predominantly owned by private companies or individuals. Public facilities, such as municipal sewage treatment plants, are relatively easy to finance. A filthy, pollution-spewing truck with two million miles on it that is the only asset of a very modest small businessman—is another matter. It must be financed by a relatively sophisticated mix of grants and loans. And then the question arises:

What is the source of revenue for the grants? And if that is not enough, which government department handles these truck loans?

Right now, these programs do not exist—either financially or administratively. They need to be created. Hopefully, you will sit on the board that creates them. And, hopefully, you will remember at least some of the more important bits and pieces from this book to guide you in their creation.

There is one other problem that we face. It is about such concepts as *global warming, climate change*, and *greenhouse gases*—and very importantly, the role that renewable energy plays in this mix of issues.

This book focuses on the key financial principles necessary to build strategies and adopt policies to deal with the environmental challenges facing humanity.

Chapter 1, "Paying for the Fix," gives some background on how we got to where we are now. It also gives examples of the kinds of challenges we face and some of the appallingly wrong-headed policies that we have either adopted or fallen into. It also sets up what might be called the *second wave* of pollution and polluters, and describes the challenges we face from them.

Chapter 2, "The 23 Principles of Environmental Finance," sets forth just what it says: a series of 23 principles that are the foundations of a sane and coherent policy to pay for a quality environment. They will be explored in depth throughout the remainder of the book.

The next five chapters explain these 23 principles in detail.

The next set of four chapters deal with loan basics and finance basics. These are the financial tools needed to understand and work with the 23 principles. Environmental finance involves grants, all kinds of loans, tax credits, equity, loan guaranties and other types of financial incentives. It is important to know how each works and where each works best in order to design financial policies that achieve desired goals.

The balance of the book deals with the major issues that must be addressed to design effective and efficient environmental finance programs.

And, of course, we must always remember that the overarching goal is to provide the greatest environmental benefit for the largest number of people at the lowest possible cost.

1

Paying for the Fix

In London in 1952, an unlucky convergence of freak meteorology and smoke caused a "Great Smog" that killed 12,000 people. In 1969, Cleveland's Cuyahoga River caught fire. In 1976, a dioxin cloud contaminated over 30,000 people around Seveso, Italy, producing serious illness and birth defects. Two years later, the problems of the ironically named Love Canal began to unfold in Niagara Falls, New York; some 21,000 tons of toxic waste had been buried, over which a housing development had been built. The company that buried it had covered it with a clay cap. But during some city construction, the cap had been breached twice. When it rained, water flooded the clay enclosure and a toxic brew bubbled to the surface. At that time, a survey of the local homeowners' association showed that from 1974 through 1978, 56% of the children born there had birth defects.*

Disasters like these galvanized public opinion to demand protection. Over the next two decades, a series of landmark laws was passed in the United States to take on these problems: the Clean Air Act of 1970 (CAA); the Clean Water Act of 1972 (CWA); the Resource Conservation and Recovery Act of 1976 (RCRA); and the Comprehensive Environmental Response, Compensation, and Liability Act of 1980 (CERCLA), better known as the *superfund law,* and others.

When I was a boy growing up on the shores of Lake Erie, I could stand in the water up to my shoulders and look down to see the bottom. When I was in law school 15 years later, I would stand in the water up to my neck and could barely see my shoulders. Now, a generation later, I can see the bottom again. Thanks to the CWA, cities like Duluth, Chicago, Detroit, Cleveland, Buffalo, and Rochester were forced—and also given massive grants—to take drastic action to improve municipal sewage treatment. Thanks to the CWA, now, if you are thirsty while sailing on the Great Lakes, you can fearlessly put a cup over the side and take a drink.

As the United Nations Environment Program noted in its quadrennial assessment of the global environment, GEO-4 (2007), we did a good job of addressing the straightforward environmental threats of the past by doing such things as building massive sewage treatment plants, requiring gasoline mileage minimums and catalytic converters on cars, and restricting industrial discharges that pollute the air and water. In the past 40 years, we picked

* In 2002, these findings were contradicted by a New York Health Department Study that concluded that chromosome damage averaged only 3%, which was only marginally higher than regional averages of 2%. But, by this time, the panic was over. See: http://www.health.ny.gov/environmental/investigations/love_canal/902news.htm.

all the low-hanging fruit. The air is cleaner than it was 40 years ago. So are the nation's waterways.

So, it looks like we won.

But if we won, why are they closing the salmon fisheries in the Pacific Northwest? Why is the *dead zone* in the Gulf of Mexico still spreading? Why are there burn bans for *indoor* fireplaces in California and Washington State?

We are now facing the second wave of environmental problems, which are tough to fix and even tougher to pay for. Massive grants and strict enforcement were the policies used to address the first wave. The second wave requires a finer hand. Enter the concept of environmental finance.

What Is Environmental Finance?

Environmental finance differs from traditional finance in that traditional finance focuses on maximizing return on investment, whereas environmental finance focuses on producing the greatest environmental benefit for the largest number of people at the lowest possible cost. Remember these words, as they are a recurring mantra: *the greatest environmental good for the largest number of people at the lowest possible cost.*

In environmental finance, there are two largely complementary approaches: (1) traditional project finance and (2) market mechanisms. With limited exception, capital improvement projects for all environmental media are financed through traditional project finance methodologies—raising debt and equity—with the important twist that lowest cost, not maximum return, is the goal.

In the field of air quality and, more recently, in water quality, there is an additional mechanism: the free market. This market mechanism is based on *cap-and-trade* programs. The total amount of pollutants that may be emitted by eligible participants is "capped" by government at a level sufficient to achieve specific policy goals. *Allowances* and *offsets* are then created and either purchased from the government or allocated by the government to polluters.

There are two financial effects of this mechanism. First, the proceeds of the allowance sales reduce the cost of the seller's (Polluter A's) pollution reduction project. And, second, it reduces the buyer's (Polluter B's) cost of compliance. So, the total cost of pollution reduction with Polluters A and B trading credits is lower than the cost of each undertaking its own projects. Recall that the name of the environmental finance game is driving down the cost of environmental improvement projects.

Note that there is an important lesson to be learned from the principle of providing the greatest environmental good for the largest number of people at the lowest possible cost. That lesson is: The lower the cost of creating an environmental benefit, the more environmental benefits will be created. If the cost of purchasing credits, or allowances, is low, then more environmental benefits will ensue. In parallel, if the cost of undertaking an environmental improvement project is low, more such projects will be undertaken.

As such, to accomplish environmental quality improvement goals, environmental finance must (1) in the case of cap-and-trade programs, increase the cost of allowances; and (2) in more traditional project finance, reduce the cost of environmental improvement projects.

The Second Wave

There is an entire paradigm shift from traditional pollution abatement to the second wave of pollution reduction. The structuring of effective environmental finance transactions has to adapt in order to address non-point-source projects (i.e., with many borrowers) rather than conventional point-source transactions (with only one borrower). For instance, traditional water and wastewater projects are point-source. They are usually large, publicly owned projects. They are few and infrequent. In any given county, there is unlikely to be more than one major sewer project per decade.

These large, public projects are easy to finance. Public bodies, like cities and counties, generally have good credit. Most of them have *investment-grade* credit ratings awarded by one of the three international credit rating agencies. The projects usually involve several millions of dollars; so they lend themselves to municipal bond financing. That is, the county sewer authority issues a bond to pay for the project. Done.

Not so easy with the second wave.

Thirty years ago, the major source of pollution in large water bodies like the Chesapeake Bay was municipal waste. Now, 30 years later, municipal waste contributes less to pollution than agricultural non-point-source transactions do. Agricultural run-off comes from a plethora of sources, but most of it stems from animal waste and fertilizer. Wherever chickens, pigs, and cattle feed, there are prodigious amounts of manure, all of which bleed nutrients like nitrogen and phosphorus into the water table, which eventually seep into and contaminate nearby bodies of water. The same is true of fertilizer, which contains the same nutrients. Unless farmers apply precisely the right amounts of fertilizer, the excess will get into the water table and contaminate it.

There are strategies for dealing with these issues that can cost anywhere from a few thousand dollars for a buffer line of trees to soak up nutrients or fencing along a stream to keep cattle out, to a few hundred thousand dollars for an Animal Waste Management System (AWMS). In comparison, however, to a multimillion dollar sewage treatment plant, these projects are tiny. In addition, farmers—even giant corporate farmers—seldom have credit ratings equal to the county sewer authority.

So, from the finance point of view, with a sewage treatment plant you have one large project, with one borrower with excellent credit. With an AMWS and stream buffers, on the other hand, you have perhaps hundreds or even thousands of individual projects, all of which are very small and all of which have borrowers with relatively inferior credit. What to do?

This problem is even more apparent in the air pollution field with small emitters. Take the case of what are called *drayage trucks*. These are the trucks that move freight containers back and forth the few miles between ports and adjacent intermodal terminals. These trucks are the bottom link of the truck food chain. They are old and filthy. They are the worst polluters on the road. And, they cluster around the already foul air at port facilities and burden the lungs of all living things downwind.

These trucks are owned by their drivers. They are the quintessential small business; and the drivers do not make a lot of money. For this reason, they can only afford the oldest, dirtiest trucks on the market, which may have upwards of two million miles on them and gone through three or four previous owners.

How to get rid of dirty, old trucks and replace them with new, environmentally friendly ones? Again, as with the farmers, you have many small projects with thousands of small borrowers with relatively inferior credit. In addition, with the dray truckers, you know—from the onset—that they cannot afford to buy new ones.

These are second-wave fixes, and they are difficult to pay for.

Finally, borrowing a couple of characters from a John Cougar Mellencamp song, we can look at another second-wave problem. It is the story of "Jack and Diane, two American kids doing the best they can." Jack and Diane live out West in a double-wide. Jack does seasonal work 40 hours a week, when there is work; but there is only work about 30 weeks a year. Jack gets paid $15 an hour. So, he brings home $18,000 gross.

Diane works 20 hours a week as a checker at the local supermarket. Her mom takes care of the kids while she is at work. She earns the minimum wage of $7.25 an hour. So, Diane brings home $145 a week, or $7,250 a year.

Together, Jack and Diane gross $25,250 a year. With their two kids, they are just above the poverty level.

In the off-season, Jack shoots a couple of deer and some ducks and geese. He catches a lot of fish. He also tends a prolific vegetable garden, all of which helps with the budget. Most importantly, he has a permit from the National Forest Service to harvest fallen timber. Jack and Diane heat their home with an old wood stove; it is their only source of heat.

The old wood stove is a menace. Their little girl has asthma and is always coughing and wheezing when the stove is lit. The smoke from theirs and their neighbors' stoves hangs like the Sword of Damocles over their valley when it is cold. The whole valley is what the air scientists call a *non-attainment* area, which means it does not have healthy air.

Jack knows that the stove is slowly killing his daughter, but he is in no position to drop his VISA® card on the hardware store counter for a $3,500 replacement.

Jack, Diane, the dray truckers, and the farmers are all part of the second wave of polluters. What they all have in common is that they do not have the cash to run out and fix the sources of their pollution. Moreover, a bank loan for 5 to 7 years at 8% to 10% interest is not going to help much.

The problem is that none of the people in these examples are legally required to do anything. There are no environmental regulations for farmland use, for port trucks, or for wood stoves. These people do not have to do anything. They can keep on polluting.

If we want them to change their ways, we have two choices: We can force them to change by regulating, or we can give them incentives to change—or, better yet, both.

Unfortunately, there are some people who will never change their ways—no matter what the effect—and believe they have a God-given, constitutionally protected, sacred right to do whatever they want with their own property, and damn the consequences. This is, of course, rubbish. In law school they teach an ancient common law doctrine with the delightfully memorable name of: *Sic utere tuo ut alienum non laedas,* which, roughly translated, means, "You can do what you want with your stuff as long as you don't harm anyone else's." This idea has been around the English-speaking world for centuries, but you will never convince some people when it comes to *their* home or *their* farm or *their* truck.

But, whether we regulate or not, we need to design incentives—financial incentives—to attack the second wave of pollution. This means that we have to give them money—our money—to get them to do what they should do anyway. That may sound crass. But look at Jack and Diane and the dray trucker. How far can we push them without offering some financial relief?

The first question that comes up whenever people talk about paying for things—especially nebulous things like climate change—is: Can we afford it? The answer is yes, although we certainly must be clever about getting our hands on the money.

Before the financial collapse of 2008, all the shares on all the stock exchanges in the world had a value of about $70 trillion. After dropping to about $40 trillion, it is slowly crawling back. In 2012, it had crept back to $54.6 trillion. In addition, according to the Bank for International Settlements, all the outstanding bonds in the world are worth about $82 trillion. So, the total wealth of the planet is approaching $150 trillion. There is plenty of money.

In addition, in 2012, the Gross Domestic Product (GDP) of all the countries on the planet was over $70 trillion per year. The British government commissioned a study that concluded that the annual cost of climate change would be $500 billion per year. This is about 0.7% of the GDP. Not a happy thought to put almost 1% of everything we do into greenhouse gas mitigation; however, in the end, it is affordable.

The United States Environmental Protection Agency (EPA) has, since 1988, run what is now a $100+ billion loan program for water pollution called the

Clean Water State Revolving Fund (CWSRF). Since 1996, they have also run what is now a $15+ billion program for safe drinking water, the Drinking Water State Revolving Fund (DWSRF). That said, you should also know that—other than some pocket change for demonstration grants—the EPA has absolutely no money for air pollution.

Here is another thing you should know: It is about the power of these two SRFs in terms of their *leverage*.

When people hear words like *leverage* in the context of money, they think it is some black art practiced by Wall Street charlatans. It is not. Everyone uses leverage; they just do not call it that.

Before the infamous days of "sub-prime" mortgages, people were required to make down payments on the homes they bought. If they bought a $250,000 home, they might be required to put down $50,000, or at least $25,000. The ratio of the purchase price to the down payment is the leverage. If the down payment is $25,000, the leverage is 10:1. If the down payment is $50,000, the leverage is 5:1. The same is true when you buy a car. If you put down $2,000 on a $20,000 car, you have achieved 10:1 leverage.

That said, the combined net assets (appropriated state and federal dollars plus interest earnings) of the EPA's two SRFs are over $40 billion. Now, governments have far better credit ratings than individuals. So, whereas a person might get 10:1 leverage on a house or car, according to the international credit rating agency, Standard & Poor's, a government finance program might achieve 75:1 leverage with conservatively run programs such as water and sewer. This means that the $40 billion of SRF assets could be leveraged into $3 trillion of environmental projects. That is far more money than is needed to pay for all the water and sewer projects on the planet, much less in the United States.

So, if the SRFs have a conservative $3 trillion of financing capacity, why does the EPA have only nickels for air pollution? It is because both SRFs (with minor exceptions) can only be used for drinking water or sewage. They cannot be used for air pollution, solid waste, or any other kind of pollution. One of the EPA's own studies concluded that the overall need for wastewater projects was only $224 billion, a sum that the CWSRF could deliver easily, with plenty of financial capacity left over for other types of environmental projects. But, despite the fact that there is $3 trillion of financial capacity in those funds, the money cannot be used for air pollution, which gets peanuts.

Why? One reason: In the House of Representatives, the Committee on Transportation and Infrastructure has jurisdiction over the CWA, while the Committee on Energy and Commerce has jurisdiction over the CAA. If only those two committees could get their Acts together! Just think: $3 *trillion* to deal with our environmental problems! And the politicians cannot put it together. Do we have the money? We sure do.

The kinsman of leverage is *term*, that is, the number of years you have to pay for something. If I lend you $100—interest free—and give you 5 years to repay me, you must pay $20 a year. If I give you 10 years, it only costs you $10 per year—half of your annual payment under the previous scheme. In other words, I can cut your payment in half by lengthening the term.

When we borrow, we are at the mercy of banks that get nervous the longer their money is out. As a result, they want us to pay for wood stoves, new trucks, and farm fixtures in about 5 years.

Governments do not have artificial term constraints. They borrow for terms commensurate with the service lives of what they are buying. Sewage treatment plants last 30+ years. So, governments issue 30-year bonds to pay for them. School buses last 10 years. So, school districts issue 10-year bonds to pay for them.

A new diesel truck will last 10 years. An AWMS will last 30 years, as will a new wood stove. Recall that Jack and Diane cannot afford to drop their Visa card on the hardware store counter and walk away with a new $3,500 wood stove. But if they could finance it through their county government, they could pay less than $19 a month. Not a pleasant thought on a tight budget, but it would certainly relieve their baby's asthma. (Furthermore, it might also help persuade them if the county passed an ordinance banning non-EPA-certified wood stoves.)

The point here is that if governments could borrow, instead of individuals, they could do so at much longer terms. This would drastically reduce the cost for the individuals. There is now a concept that does just that.

In 2008, a new trend began to sweep the country that would let Jack and Diane pay for their new wood stove over 30 years. The EPA's Environmental Financial Advisory Board (EFAB) developed "VEIBs," or Voluntary Environmental Improvement Bonds. The United States Department of Energy (DoE) calls the entire effort the "PACE" program, which stands for Property Assessed Clean Energy. The DoE won the publicity battle, so most adhere to the latter designation, the PACE program.

This all started in Berkeley, California, when Mayor Tom Bates wanted to encourage residents to reduce the carbon footprint of their homes by installing solar panels. These can cost $20,000 to $40,000. With a 5-year bank loan at 8%, homeowners would pay $417 to $835 a month.* Furthermore, when they sold their home, they would have to pay off the balance but leave the solar panels for the new owner. Not a happy prospect.

So, Mayor Bates devised a plan whereby the City would buy the solar panels for any homeowner who wanted to join the program. The City could issue a municipal bond to pay for the solar panels and get repaid by placing a tax assessment on the homeowner's property. Solar panels have a service life of 20 years. So the City issued 20-year bonds. This reduced payments to between $134 and $267 per month. More importantly, when the homeowner

* If you are checking my math, you will think that this is inaccurate. Not so. Municipal bonds have annual principal and semi-annual interest payments. You sum them and divide by 12 to get a monthly payment. This produces a slightly different result than, say, a traditional mortgage that actually has monthly payments.

sold his home, both the tax lien and the solar panels stayed with the property. So, the new owners—who got the benefit of the solar panels—also paid their fair share. In its first year, the Berkeley program was oversubscribed.

As a "charter city" in California, Berkeley had the power to do this. But many other jurisdictions did not have these powers. Accordingly, the State of California passed a new law enabling all local governments in California to do likewise. So did the State of Colorado. More than a dozen communities in both states began these PACE programs, and other states started to pass similar legislation.

What the EFAB pointed out in its 2009 report was that, as great an idea as this is, these states and local governments are missing a much bigger boat. The EFAB noted that these VEIBs could be used to finance a host of environmental improvements, not just for energy efficiency but for air pollution and non-point-source water pollution as well. The VEIBs can be used to finance AWMSs on farms, and they can finance Jack and Diane's new wood stove.

Here are some of the environmental improvements that can be financed with VEIBs:

- Solar panels
- Insulation
- New insulating windows and doors
- Tankless water heaters
- Geothermal loops
- The replacement of old wood stoves and hydronic heaters with EPA-approved devices
- Permeable pavement
- Rain gardens
- Rainwater management systems
- Green roofs
- Replacement of failing septic systems
- Animal feeding operations
- Animal Waste Management Systems
- Structures for stream crossings
- Stream buffers (trees and fences)
- Replacement of agricultural diesel equipment
- Lead paint removal
- Asbestos removal
- Radon mitigation
- Indoor plumbing
- Resiliency retrofits for buildings in high hurricane or tornado areas

The Berkeley program is truly a stroke of genius. First, it is totally voluntary, so nobody's taxes get raised unless they want to. Second, both the improvement and the assessment go with the property, so successive owners both get the benefit and pay their fair share. And third, it extends the *term* of repayment far enough to make it affordable for more residents.

Unfortunately, the Berkeley story does not have a happy ending.

The Federal National Mortgage Association, which we all know and love as Fannie Mae, is the largest purchaser of home mortgages in the United States. When you go to a local bank for a mortgage, they give it to you; but then they turn around—immediately—and sell it to Fannie Mae. Fannie Mae is a Government-Sponsored Enterprise (GSE). Despite the fact that its shares traded on the New York Stock Exchange for 42 years, it is regulated by the Federal Housing Finance Administration (FHFA). Once the sub-prime crisis hit the housing finance industry, the FHFA took a close look at the Berkeley program and decided that it disliked the idea of the solar panel liens taking legal precedence over mortgages owned by Fannie Mae, which was losing billions of dollars every month. So, the FHFA issued an order to Fannie Mae to stop buying mortgages with PACE liens on them. That brought the PACE program to a dead halt. Lawsuits were filed. Bills were introduced in Congress. But there the matter stands.

Where will it all wind up? Well, for probably over a hundred years, when the county ran a sewer line down a street that was on septics, it required the homes to hook up to the sewer system. It then proceeded to connect each home and after it did so, it slapped a lien on that property—just like a PACE lien. The homeowners were given usually 20 years or so to pay for the connection. The FHFA knew that this was going on everyday all across the country, but they never objected. It still is, and they still have not objected. They only object to the PACE lien. Why?

First of all, the solar arrays cost $20,000 to $40,000, which is far more than the average sewer connection costs. Second, the FHFA did not like the discretionary nature of the PACE program. When it comes to sewers, the homeowners have no choice: They must connect to the sewer system. Yet with solar panels, it is up to the individual homeowner to decide.

We will be interested to see where the FHFA goes with *resilience*. Resilience is the new climate change buzzword that relates to the ability of a building to withstand severe weather events like hurricanes and tornadoes. It costs about $150,000 per house to retrofit them by putting them on stilts to withstand high water caused by flooding. These retrofits are not electives. If you do not retrofit, you cannot get flood insurance. If you do not get flood insurance, you cannot get a mortgage.

Many people will need help paying for these retrofits. Some will not be able to afford them. A natural thing for a local government to do would be to undertake a PACE-like program to help homeowners pay for these retrofits. Expensive? Yes. Discretionary? No. So, let us see where the FHFA will go with this new wrinkle in the PACE program.

Thus, somewhere in this mélange of intrigue lies the eventual solution to the problem, whether it comes from Congress or the courts. If Jack and Diane's county banned old wood stoves and created a PACE-like incentive loan program for those (1) in non-attainment areas, (2) with aged or infant residents with pulmonary disease, and (3) for households close to the poverty level, I doubt the FHFA would touch it. They would not have a legal leg to stand on. The same is probably true with the resilience retrofits.

Looking back to the list of environmental improvement projects that can be financed by VEIBs (above), the items that directly relate to public health, such as wood stoves, radon mitigation, and indoor plumbing, can probably succeed using PACE liens. Some of the others—like stream buffers or insulation—might not.

One of the reasons the PACE program was so attractive was that it lowered the payments on solar panels by lengthening the term over which they were paid for. As previously mentioned *term* is one of the most important concepts in environmental finance. Another one is *leverage*. Both are included in the following example.

In 2004, the Maryland General Assembly passed what they called the Bay Restoration Fund Act, which levied a $2.50-a-month fee on every dwelling served by a major sewer system *and*—for the first time in history—the same fee on the owner of a septic system. In 2012, they doubled this fee to $5 per month. The money from the septics goes into a separate account that is used for non-point-source pollution programs*.

The money accumulated from users of the sewer system goes to pay for "Enhanced Nutrient Removal" (ENR) at the State's 66 largest sewage treatment works. Maryland originally estimated the problem at about $800 million. They figured their "fee" on sewer users would bring in about $60 million a year, which they could pledge to the repayment of the $800 million of bonds that they would issue to fund the ENR program.

Now, here is how this works. If the only mortgage you could get was with a 15-year term and you could only afford to pay $1,500 a month, then (at a 5% interest rate) you could only afford a $190,000 mortgage. But, if you found a different bank that would give you a 30-year mortgage, then you—with the same $1,500 a month—could get a $280,000 mortgage. In other words, you could get $90,000 more home for the same monthly payment.

Maryland tried to do this but ran into two problems. First, their estimates for fixing the problem rose from $800 million to $1.3 billion. Second, they realized that the State Constitution limited their bond term to 15 years. That left them with about $670 million in financial capacity to pay for a $1.3 billion problem.

The solution to Maryland's problem is a one-word amendment to the State Constitution: Change *15* to *30*. Is this prudent? Yes, because the service life of these ENR plants is 30 years; so they should be financed for 30 years. The US

* The people of Maryland pleasantly refer to this as the *flush tax*.

Department of Agriculture has been financing rural water and wastewater systems for terms of *40* years since 1941 without any problems. How many members of the Maryland General Assembly do you think have 30-year (versus 15-year) mortgages on their homes? However, despite the fact that the State of Maryland and the EPA have been sued by the Chesapeake Bay Foundation for their lax stewardship of the Bay, policymakers have yet to amend the Maryland State Constitution to finance projects over a longer term.

The Maryland story is about wasting badly needed money. So, too, is the story about Congress' not extending the $40+ billion SRFs to cover all environmental hazards. Here is another one.

In the $870-billion American Reinvestment and Recovery Act (ARRA) of 2009, there is a provision that grants up to $1,500 tax credits to those who replace their old wood stoves. In Jack and Diane's smoke-covered valley, there is a weekend getaway cabin owned by some prosperous folks from the city. Now, the $1,500 tax credit might persuade the city couple to replace their stove, but Jack and Diane would not be persuaded by the ARRA tax credit. With their income, Jack and Diane barely pay any taxes at all, so the credit is of little real value to them. Also, the hassle with the tax credit is another hindrance. Furthermore, and most importantly, dropping the price of the new stove from $3,500 to $2,000 does nothing for them. They cannot afford $2,000 either.

Here is a better way of dealing with this problem:

Instead of a $1,500 credit for rich and poor alike, Congress could have appropriated the same amount of money that the credit would cost and given it to the states for incentive grant programs, especially if those states would come up with some sort of matching funds. Then, if the states were smart, they would target the money.

States could first target the money to the problem. They could limit their grants to non-attainment areas—like Jack and Diane's valley—where the fine particles of soot that old stoves generate are the real cause of the problem. Second, states could target the grants to those—like Jack and Diane—with low income levels *and* who had no other source of heat in their homes, *and even further* to those homes where people were suffering from asthma or other pulmonary diseases aggravated by the airborne soot. In short, the $1,500 credits that Congress enacted need not have been wasted on the wealthy and the well. If Congress had excluded all those who live in attainment areas and can afford to pay for new stoves, there might have been enough left to give Jack and Diane a 100% grant, which would have meant a healthier valley and a healthier baby.

The lesson on financial incentives for environmental improvements is that the less expensive the improvement, the more people will take advantage of it. If you can reduce the cost of solar panels, or new dray trucks or wood stoves, more people will buy them. That is the job government can do and should do: Lower costs for the people.

A word about interest rates. Governments can borrow at lower rates than individuals. This is because the average state or local government has a better credit rating than do most individuals. Jack and Diane pay 18.9% on their

Visa card. Second mortgage lenders might charge 8% to 9% for a second mortgage loan for solar panels. A finance company might charge 12% to 15% for a new dray truck. Governments can borrow—in the same market—for about 5%. So, if they can pass this reduced rate along to their people, *their* costs will be lower too. And, the lower the cost, the more takers there will be with any environmental incentive program.

When we say that governments—instead of individuals—should be doing the borrowing, there should be a little fire bell going off in the back of your mind. Everybody knows about credit ratings. If you have 700+, you get a good rate; if you have a 500 rating, you either get an astronomically high interest rate or no loan at all. These ratings are based on models of default. The lower your credit rating, supposedly, the higher the probability is that you will default on your loan. So, if the government interposes itself between a low-rated borrower and the credit market, is the government not likely to suffer a lot of defaults? And, if so, where will it get the money to make up for them?

There are ways of dealing with this problem. In both Jack and Diane's case as well as that of the farmer, a VEIB program can mitigate their credit risk by attaching the payment obligation to their home or farm via a tax assessment. In the case of the dray truckers, their problem can be dealt with by aggregation and buffering.

Aggregating bad credits sounds like a bad idea. There is an old saying that you should not try to make chicken salad out of chicken manure. A hundred bad credits do not make a good credit. That is why buffering is also an essential component.

There are two strategies to buffer a dray truck loan program. The first is to make it a *lease* program, not a *loan* program. If a trucker fails to make lease payments, the truck can be easily repossessed and re-leased to another trucker. If it is a loan program, the lender needs to go to court—a longer and more expensive process. So, a lease program means that if one trucker defaults, another will take his place swiftly—and resume payments on the lease.

The second is *overcollateralization*, or Self-Funded Reserves (SFRs). If you have 100 truckers in a lease program, not all of them will have problems. Some bank statistics suggest that fewer than 5% will have payment problems. Of these, most will either resume payments, or they will lose their truck and another driver will assume the lease and resume payments. So, out of 100 truck leases, there will probably be problems with 5 of them, or fewer.

Now, here is how "overcollateralization" or SFRs work. If each truck costs \$100,000, you borrow \$110,000—not just \$100,000—and put the additional \$10,000 into a common (self-funded) reserve as collateral for all 100 trucks. So, for 100 trucks at \$100,000 apiece, you have a \$10,000,000 program that can be financed through municipal bonds issued by a government agency, such as the Port Authorities where the drivers work. But instead of issuing \$10 million of bonds, the Port Authority issues \$11 million and salts away the additional million into a reserve account in case any of the truckers miss their lease payments. More about SFRs and overcollateralization in Chapter 18 dealing with Credit Enhancement.

But wait! Have we just created $1 million out of thin air? Who pays for the extra million? The answer is that the truckers pay for it, in the same way you pay for "points" when you get a mortgage.

When you get a $300,000 mortgage "plus two points," you are actually getting a $306,000 mortgage; that is, the bank tacks on the 2% to the mortgage principal. So, too, with the dray truckers; the extra $10,000 gets added onto their $100,000 loan. But, in their case, it is not 2%; it is 10%. While that is a bit pricey, especially given that dray truckers are pretty far down on the income scale, there are a few considerations to remember. First, most dray truckers cannot get any financing for the third- or fourth-generation trucks they buy with two million miles on them. They have to pay cash. Second, by using government bonds, they can finance brand-new trucks for 20 years. Because of the short distances they drive, a dray truck is likely to only put 40,000 miles a year on its odometer. Over the 20-year period, that is 800,000 miles. So, as it turns out, at the end of their 20-year lease, the average dray truck in the leasing program is likely to have 1.2 million *fewer* miles on it than the average dray truck today.

A $100,000 loan for 20 years at, say, 5% requires a payment of $670 per month.* Add $10,000 to the loan, and the payment only rises to $712 per month. Pleasant, no; doable, yes.

There are several other reasons—not strictly financial—why a dray truck-leasing program at a port is a good idea. First, when a Port Authority buys 100 trucks, it will get a volume discount—probably 10% or more—which wipes out the *overcollateralization* cost. Second, the participants in the program can constitute a "fleet" for the purposes of getting "fleet fuel discounts." This will knock off 5% to 6% of their fuel bills. Third, going back to the credit problem, if the Port Authority is really worried about getting paid by the dray truckers, they can simply estimate the number of containers the truckers haul each month, and, by arrangement with the shippers, deduct a certain amount from each container payment to pay for the monthly lease obligation. Finally, if they are still worried, they can adopt a regulation prohibiting entry to the port for any trucker more than 60 days in arrears of his lease payments. So, as you can see, there are a lot of clever ways to deal with the problem of multitudes of small, inferior credits.

This port-leasing scheme might possibly happen. The Ports of Long Beach and Los Angeles have begun a program where each container entering the port is surcharged and the proceeds used for grants to replace old dray trucks. They are pretty unique. They are the two largest ports in the country and the winds from the west blow all the pollution they generate into Los Angeles and Orange County. So, under relentless—and well-deserved—pressure from the California Air Resources Board (CARB), they have heeded a clear wake-up

---

* See footnote page 7 on payment calculations.

call. There are also signs of life in Oakland and in Seattle/Tacoma. But the rest of the US ports seem to have their heads planted firmly in the sands of denial. Lease trucks? Oh, we don't do that! We're a port!

Ambassador and, later Governor of New York, Averell Harriman, whose father was probably the biggest railroad baron in US history, once said that most of the railroads his family owned went out of business shortly after they sold them. He said the reason for this was that the old management teams believed they were in the railroad business; they did not realize they were in the transportation business.

Ports used to think that they were just docks. Then they realized that some ports offered advantages that others did not have. Consequently, they then got into the economic development business. As the iron wheels of government crank out ever stricter air quality regulations, the ports—whether they like it or not—may realize that they are in the air pollution control—and therefore the truck leasing—business.

Therefore, strategies can be devised to deal with two of the major issues facing the second wave of pollution—many small sources with relatively inferior credit. But there is a third, much larger problem imbedded in this second wave.

The CWA deals with both point- and non-point sources of water pollution. For point-sources, think of an effluent pipe coming out of a factory or a municipal sewage treatment plant. For non-point, think of how brown rural streams get after heavy rains. The brown is soil that has washed away, much of it from farmland. With the brown sediment comes nitrogen and phosphorus from fertilizers and animal waste.

Recall that the Clean Water SRF is $100+ billion in size. It can finance both point- and non-point-source pollution control projects. Does it surprise you to learn that 96% of the $100+ billion has been spent on point-source projects? Only 4% has been spent on non-point-source projects. The reason for this is simple. Point-sources are highly regulated. Every effluent pipe is subject to a National Pollutant Discharge Elimination System (NPDES) permit that spells out precisely what pollutants can be discharged and how much of each. If your discharge pipe puts out more pollutants than your permit calls for, then you must undertake a project to fix it. If you are publicly owned, like municipal sewage treatment plants, you can borrow money from the SRF to fix your problem. If you are privately owned, you cannot. Nonetheless—public or private—you are *required* to fix it. If not, you pay substantial fines and can even face criminal charges.

Contrast this with a farm with hundreds of acres of crops, all of which need fertilizers, pesticides, and herbicides, and/or hundreds of head of livestock that produce tons of waste each week. Much of the fertilizers, chemicals, and manure winds up in the nearest water body and pollutes it just as surely as any effluent from an NPDES-permitted facility. However, there are no NPDES permits for non-point-sources. There are no permits at all.

Controlling the run-off from farms, or any other non-point-source of water pollution, is totally voluntary.

Now, knowing that controlling non-point-source water pollution is totally voluntary, does it still surprise you that only 4% of the CWSRF money has been spent on non-point-source projects? Does it surprise you that millions of farmers and others have not *volunteered* to spend huge amounts of their personal funds to fix their pollution problems? This becomes even clearer when you look at stormwater. One of the major problems with stormwater is the effect of impermeable pavement. So, say you are a county commissioner and you want to reduce the effects of impermeable pavement. You can, of course, ban impermeable pavement. If you do, however, you can measure your future political career in nanoseconds. You can go to the state SRF and organize low-cost loans, even zero-interest loans. But how do you persuade the owner of a small shopping center with, say, a 1-acre parking lot to rip up the pavement and replace it with a permeable surface? Replacing asphalt costs about $10 per square foot. So, that is $436,000 that the owner is expected to borrow and pay back out of his own funds. Not likely.

Green roofs, too, mitigate stormwater effects. They also cost about $10 a square foot. So, Mr. or Ms. Commissioner, you want the owners of flat-topped buildings to install green roofs? Are you going to try to force them? Not if you like your job. How are you going to persuade the owner of a building with a flat 100- × 100-foot roof to borrow $100,000 to install 10,000 square feet of green roof and pay it back out of his personal funds? You are not going to. This is the really big problem with the second wave of pollution and polluters.

The answer to this dilemma is that you have to pay them to do it. You have to pay the parking lot owner to replace his asphalt; and you have to pay building owners to plant green roofs.

A couple years ago, a Washington, DC lobbyist for a major environmental NGO once confided that he had a defective septic system on his property that was very close to a very sensitive stream. When asked if he was interested in a low-interest $15,000 loan to replace it, he laughed. When asked if he would do it for a 100% grant, he thought for a minute and said no. He said it was not worth the hassle of having his life disturbed and his yard torn up for a year. When asked if he would accept a 100% grant plus another $2,500 for his trouble—a 117% grant—he reluctantly said yes. So, there we are. And this gentleman is an environmental lobbyist!

Fixing the second wave of pollution requires taking on thousands and thousands of small projects. Almost every one with inferior credit—assuming we can get them to borrow. That, of course, is the biggest hurdle of all that we face: finding the money to induce people to undertake *voluntary* projects.

The past few pages have introduced you—through stories, anecdotes, and examples—to the role that finance plays in the environmental improvement game. The next chapter of this book deals with "The 23 principles of environmental finance." These are the principles we need to use to build effective strategies to deal with these second wave issues.

2

The 23 Principles of Environmental Finance

This chapter sets forth a series of principles to guide the development of effective finance policy to improve environmental quality.

Two of these principles are unique to environmental finance. A few are simply a recitation of a Best Management Practice (BMP) in business. Several are principles of the basic mathematics of finance. Others are common sense. Regardless of provenance, they bear restatement. And they bear restatement together, in one place. The instances in which these principles occur are not in isolation. Simple principles, such as *putting just one person in charge*, occur in tandem with much more complex concepts, such as *wherever possible, use loan guaranties instead of loans*. Ideally, these principles should be condensed to a single page and kept in the center desk drawer.

To ease the use of the principles of environmental finance, I present all of them in one chapter. As you will see, I have broken them up into five categories. First are the *two core principles* of environmental finance. Next are the *policy principles* because it is logical to begin by defining policy objectives. Third, the *fundraising principles* guide the use of enacting revenue-raising measures to finance environmental programs. Fourth are the *financial principles,* some of which may shock you, or at least seriously surprise you. Last are the *management principles,* none of which are unique to environmental finance and some of which you may be familiar with.

Some of these individual principles will only be discussed briefly because that is all they need. Others will be the subjects of entire chapters of their own. *Leverage,* which again is arguably the most important financial principle, really deserves a book of its own.

Finally, this list of principles was developed over about 10 years. For the first few years, there were 12 principles. That slowly crept up to 19, which stuck for 3 years. Then in one semester, the last four were added. The point here is that this list is a living thing. If you disagree with any of the principles, or think there are more than 23, contact me: mcurley1@jhu.edu. Your contribution—pro or con—will be most welcome.

Two Core Principles

1. The fundamental goal of environmental finance is to deliver the greatest amount of environmental benefits to the largest number of people at the lowest possible cost.
2. The way to achieve this goal is to drive down the cost of environmental projects. As such, the principal aim of environmental finance is to decrease the cost of environmental projects.

No improvement in environmental quality—unless by act of God—is possible without spending money. A high-quality environment costs money—serious money. This money will come right out of every citizen's pocket. Even though they may prize the environment, they are not going to be happy about paying for it. An example from Maryland can illustrate why these two core principles are the most important.

Right now, every household in Maryland pays a $60-per-year Bay Restoration Fee that, because it refers to sewage, the press has dubbed the "flush tax."* Add to that the cost of stormwater remediation, which is conservatively estimated at over $1,000 for every man, woman, and child in the State. If that cost were financed with a long-term, low-interest rate municipal bond, the Annual Debt Service Payment (ADSP) would be about $60 per year. But this is not just another $60 per household; this is $60 per *individual*. With about 2.5 people per household, that is another $150. So, now the household cost is up to $210!

People in Maryland are getting angry. In 2012, the State mandated the nine largest counties and the City of Baltimore to create stormwater utilities *and to fund them with a stormwater tax or fee*. To show their contempt for the State's mandate, one of the nine counties passed a $0.01 tax per parcel of land. This tax will raise less than $500—that's total, for the whole county. This is clearly a message to the Governor and the General Assembly. Both will have to deal with this issue during the 2014 legislative session.

So, if we really want a quality environment, we must provide the greatest amount of benefit for the largest number of people at the lowest possible cost. Our overarching policy must be to drive down the cost of environmental improvements. Chapter 3 is entirely devoted to these two key principles.

Policy Principles

3. Regulation and financial incentives are complementary policies. Carrots and sticks work best together. They should be used in tandem.

* Much of which is wasted, as you will see.

4. Forget about divisive canards like "the polluter pays." We are all polluters, both directly and indirectly. We are all going to have to pay. And the more we pollute, the more we should pay.

The Policy Principles are presented in Chapter 4.

Management Principles

5. Put one person in charge. Make that person accountable, and be prepared to fire him/her.
6. Advisory committees work; executive committees do not.
7. Make strategies and programs fully transparent (i.e., by posting transactions on the Internet). Doing so will enhance the legitimacy and accountability of environmental programs. If people are going to pay, they need to know their money is wisely spent.
8. Be flexible. If one strategy does not work, get rid of it and adopt another.
9. Develop new strategies. Issues like non-point-source water pollution and small-source air pollution cannot be tackled with today's legal and administrative tools. New ones must be designed and adopted.
10. The time to implement the tools of environmental finance is now. Everything costs more in the future as a result of inflation.

The Management Principles are presented in Chapter 5.

Fundraising Principles

11. Raise money from many small charges, fees, or taxes—not one big one. Many small sources of money are more stable than one large one.
12. Once collected, put all the environmental money in one basket. Do not fragment or piddle it away.
13. Change behavior while raising money. Do not tax all equally; tax the polluters more while rewarding energy efficiency and green practices.
14. Use *dedicated revenue streams* (such as annual taxes or fees) to finance capital, not operational, expenses.
15. Make it as painless as possible. At the federal level, raising the rates on a general tax—like the income tax—will set off howls of protest across the country. It will mobilize armies of lobbyists in

Washington who will roam the halls of Congress persuading the members to oppose it. At the state level, the same will occur if there is an income tax increase proposal or a real property tax increase proposal. Virtually every newspaper in the state will editorialize against it. And the lobbyists will mob the state legislature. The more opposition, the less likely that an effective environmental finance program will be adopted.

The Fundraising Principles are presented in Chapter 6.

Financial Principles

16. Leverage the money. The most important word in improving environmental quality is *leverage.*
17. Stop general subsidies, which waste billions of dollars. Target subsidies only for those who need them.
18. Make financial decisions based on strict cost/benefit analyses. Get politics and chance out of the decision-making matrix.
19. Never give grants unless absolutely necessary.
20. Never make loans; guaranty them.
21. Finance assets over the full term of their service lives.
22. Insist on full-cost pricing of environmental services. Full-cost pricing will drive technological innovation. In turn, new technologies will drive down costs.
23. Use good science and good economics to get accurate cost estimates. Low estimates cause anger and frustration; high ones generate denial.

The Financial Principles are presented in Chapter 8.

As noted before, these principles are really guidelines for the development of sound and effective finance policies for improving the environment.

As stated, no environmental improvement—unless by act of God—is possible without spending money. Much attention has been given in government, academia, and even the media to the development of environmental policy. But little or no concern is evinced in these quarters for where the money will come from to implement such policy or, more importantly, how that money will be spent.

Several years ago, the British Exchequer commissioned Sir Nicholas Stern to estimate the cost of dealing with climate change. Sir Nicholas estimated that it would cost $500 billion a year. That looks like a staggering sum! But in the perspective of global GDP (Gross Domestic Product) of $50 trillion, the estimated cost of addressing climate change only amounts to 1%. Nonetheless, no matter how it stacks up to global GDP, $500 billion a year is, indeed, a very substantial amount of money. Therefore, it is imperative that—as we develop the policies we need to deal with environmental problems—we spend at least equal time developing financial strategies to implement those policies. First and foremost, we need to know that they will achieve the desired environmental results. Second, we need to know that the money we raise will be enough to solve the problems. And, third, we need to know that those funds will be affordable and equitable for all.

Now, let us begin to look at these 23 principles in some detail.

3

Two Core Principles of Environmental Finance

Of the 23 principles of environmental finance set forth in Chapter 2, the first two are, by far and away, the most important. You will have seen, and continue to see them repeated many times throughout this book.

Principle 1: The goal of environmental finance is to deliver the greatest amount of environmental benefits to the largest number of people at the lowest possible cost.

The goal of traditional finance is to maximize returns on investment. The goal of government finance is to minimize the cost of delivering essential public services. The goal of environmental finance is to deliver the greatest amount of environmental benefits to the largest number of people at the lowest possible cost.

This is a continuing mantra. Politicians and practitioners need to keep this simple concept squarely in front of them when devising programs to pay for the environment.

You may think that this goes without saying. Not so. I know many people who manage environmental finance programs all across the country. You would be very surprised at how many of these people believe that their principal responsibility is "not to lose money." They believe that their primary goal is to preserve every nickel with which they have been entrusted. They act like bankers with anxious shareholders breathing down their necks. Worthy goal, you say? No. Their primary goal is to make the nation's air and water, etc., cleaner. That is the worthy goal. Husbanding money is certainly part of their responsibilities; but improving the environment and our quality of life comes first and foremost. Furthermore, this fear of losing money has an unfortunate side effect: risk aversion. Risk aversion is a plague on humanity. It is the opposite of innovation. Everyone wants *innovative finance,* yet few will practice it.

Only 4% of the $90+ billion Clean Water State Revolving Fund (SRF, Self-Funded Reserve) in the 50 states and Puerto Rico has been spent on non-point-source water pollution. There are several reasons for this sad performance, but one of the major factors is certainly risk aversion.

Eligible point-sources are always owned by local governments or their agencies.* They generally have excellent credit ratings, awarded by one or more of the three major international credit rating agencies. They are very easy to lend to. Repayment is virtually a foregone conclusion.

Non-point-sources are very, very different. Most are owned by you and me and our friendly farmers and small businesses, that is, the private sector. No matter how good your credit score is, it pales before the credit rating of your local water or sewer authority. Tomorrow you might lose your job or suffer some financial catastrophe and not be able to pay for the rain garden you installed with a loan from the SRF. However, your local water authority will have no such concern. They have enough cash reserves to get them through fiscal crises; immediately after which they will raise your rates to make sure they have enough money to repay their SRF loan. So, an SRF loan to the Local Water and Sewer Authority is almost risk-free. But, relatively speaking, an SRF loan to you or me is loaded with financial risk.

The vast majority of privately owned, non-point-source projects are funded by SRFs through what are called *loans-to-lenders programs*. In these programs, the individual SRFs do not take the credit risk at all. Rather, they fob off the credit risk onto local banks. Here is how they work: First, the SRF makes a loan to a bank at, say, 2% below what that bank would pay you on a Certificate of Deposit (CD). Then the bank makes a loan to you at a rate 2% less than they would normally lend to you. (Of course, there is no way to know at what rate the bank would normally lend to you!)

Let us say the bank's CD rate is 4%. The SRF loans to the bank at 2%. Then, if the bank would normally lend to you at 8%, but the SRF is accepting 2% less interest, the bank will do the same and lend to you at 6%. In the end, the bank is getting exactly what it wants—8%, 6% of which comes from you and 2% of which comes from the SRF.

However, here is the rub: If you do not pay back your loan, the bank gets stuck. The bank must still pay back the SRF even if you do not pay them back. Why? Because the SRF does not want to deal with your credit. The SRF is risk averse.

George Ames, at the EPA (US Environmental Protection Agency), was one of the founders of the SRF program and now manages it. He often pulls his hair out (figuratively) because many individual SRFs think they were created to make money (or, at least, not lose it) for their state. They forget that they were created to make our water clean.

The great irony of this is that ALL of the SRFs programmatically lose money every day by lending at rates that are below the rate of inflation. That is like a bank paying you 4% on a CD and making you a loan at 2%. They certainly will not be in business long with policies like that.

---

* Privately owned point sources are *ineligible* for funding under the Clean Water Act, one of its great failings.

I hope you also see that this loans-to-lender concept is a great waste of money. The SRF should have loaned you the money at 6%. Then the state would be making 6%, instead of 2%.

Do you understand how crazy this is? The SRF thinks that it is doing a good job by earning 2% with zero risk. Rubbish. If they assumed the risk and charged 6%, they would be ahead of the game as long as fewer than 4% of their borrowers defaulted. On top of that, the SRF can secure your loan with a lien on your house. Under such circumstances, it would be a "cold day in hell" for 4% of their borrowers to default. And, in the final analysis, even if you defaulted on your SRF loan, the people of your community and your state would have gotten the environmental benefit of your rain garden—which is the purpose of the Clean Water Act and the reason that SRF staff have their jobs in the first place.

As you can see, some government employees who run environmental finance programs forget Core Principle 1: that the goal of environmental finance is to bring the greatest environmental good to the largest number of people at the lowest possible cost. This forgetfulness is not limited to the SRF staff; it afflicts some of their overlords in Congress and state legislatures. Most clearly undermined is the "lowest cost" part of the mantra. Why? Because the SRFs sacrifice lowest cost for lowest risk.

Risk aversion is a human factor. As such, it is almost understandable. But there are institutional elements within laws that have the same pernicious effect: ones that raise the cost of environmental improvement projects.

The examples I would like to cite are called *crosscutters.* This can be a very sensitive subject. Let us look at a case in point.

In 2009, the US Congress passed the American Reinvestment and Recovery Act (ARRA) to help dig the country out of the sub-prime mortgage mire. Because water and wastewater projects are so labor intensive, Congress added several billion dollars to the Drinking Water SRF and the Clean Water SRF. They did the same in the next year's federal budget. But in 2010, they attached two strings. The first was a "Buy American" requirement. Assuming that local sewer agencies were doing a good job of picking cost-effective technologies to solve their problems, the "Buy American" requirement not only put a serious crimp in their plans, but also added costs. Putting a "crimp in their plans" was not just an inconvenience; serious wastewater projects often take three or more years to plan. So, throwing in a last-minute proviso to "Buy American"—after three years of work had determined that, for example, a German technology was most cost effective—wasted a huge amount of time and effort and, assuming that the analytical work was correct, produced a less cost-effective result.

Protecting American jobs seems laudable enough, but "Buy American" requirements probably are not the way to do it. Many years ago, when the State of New York was on the verge of bankruptcy, the governor told all state purchasing agents to "Buy New York" and urged local governments in New York to do the same. Immediately, the Governor of Pennsylvania announced that if New York implemented its "Buy New York" policy, that he would ban the purchase of any New York products by his purchasing agents and would urge his local governments to do likewise. There was no official reaction from the Governor's Mansion in Albany, but it was rumored that forward elements of the New York Army National Guard's 42nd Infantry Division were amassing along the Pennsylvania border.

Silly. But it makes the point about trade wars and worse: If every country adopted a "Buy-Only-Our-Own" policy, world trade would grind to a halt and the have-nots would slip further into the morass of ignorance and poverty.

For Congress to foist a "Buy American" requirement on the SRF is a special irony. They do this because they want to help American manufacturers, right? Well, guess what! Congress has excluded American manufacturers from the financial benefits of the SRF. If you are a local government and you have a National Pollutant Discharge Elimination System (NPDES) permit, you can get financial help from the SRF. But if you are an American manufacturing company with an NPDES permit—struggling to make your weekly payroll—you get nothing. So much for Congress' efforts to help American manufacturers.

Also, as if the "Buy American" problem was not enough, Congress sent yet another crosscutter to vex the SRF program in FY 2010. This one concerns the infamous Davis–Bacon Act. A common misperception of the Davis–Bacon Act is that it requires labor union-scale wages to be paid on all federal contract work; and that union-scale wages are almost double non-union-scale wages. Looks like a classic pro- versus anti-union issue. All of this, however, is not quite true.

To begin with, Davis and Bacon were not pro-union Democrats; they were Republicans. So was President Herbert Hoover, who signed their Bill into law in 1931. When enacted, it had absolutely nothing to do with labor unions. It was passed because there were huge disparities between the wages paid to construction workers across the country. So, for example, when a federal construction contract in New York was put out to bid, a contractor in Mississippi—where wages were very low—could win the bid and truck local workers up to New York and house them in shacks while on the job. This deprived local workers in New York—who had elected the congressmen and senators who got the pork-barrel project in the first place—of these valuable construction jobs.

To remedy this, the Davis–Bacon Act required that "prevailing" local wages be paid; thus neutralizing the cost competition between a Mississippi and a New York contractor. The Mississippi contractor could still truck his workers to New York, but he would have to pay them the same wages that a New York contractor would pay his workers. What is the point of that?

Apparently, in many areas, the "prevailing" wage is a union-scale wage. It is so because most of the work in that area is evidently done by union contractors. As such, these union wage scales "prevail." But it is so because of happenstance, not because of the Davis–Bacon Act.

Nonetheless, the lore is that a Davis–Bacon requirement can double the cost of labor. Now, in water/wastewater work, there is a rule-of-thumb that materials comprise 50% of the average project and labor the other 50%. So, on a \$20-million project, \$10 million would be for materials and \$10 million for labor. Now, if a Davis–Bacon requirement were saddled onto the project, the labor cost could double to \$20 million. This means a \$30-million project instead of a \$20-million project; and that your local sewer authority will have to raise your rates 50% *more* because of the Davis–Bacon requirement. This is hardly "bringing the greatest environmental benefits to the largest number of people at the lowest possible cost." It is ironic that on January 20, 2011, as part of their fiscal package to wring out \$2.5 trillion from the budget over the next 10 years, the Republicans in Congress tried to repeal Davis–Bacon, yet they voted to make it a permanent fixture of the SRF program beginning in FY 2013. Apparently they felt it was too expensive for the federal government, but not so for the states.

In any event, whether the issue is a well-meaning loans-to-lenders program or a well-meaning crosscutter like "Buy American" or "Davis–Bacon," if you accept them, then be aware: You are *not* providing "the greatest environmental benefits to the largest number of people at the lowest possible cost."*

Principle 2: Drive down the cost of environmental improvements.

This principle not only gets its own chapter (the final chapter of this book), but it also gets large chunks of three other chapters.

Let me, therefore, just give you the flavor of where we are going with this concept right now because you will see so much about it later.

Several years ago, a wealthy doctor in Chautauqua, New York, built a geothermal system to provide renewable energy to a compound of several houses he owned along the shore of Lake Chautauqua. The system cost him \$60,000, which he paid in cash. This system saved him \$1,000 a month (Yes, a month!) in electricity charges. So, using straight math, with no discounting, it took the gentleman 60 months (or 5 years) to break even, or get his money back, on the project.

Recall from Chapter 1 the discussion of a renewable energy program called the PACE (Property Assessed Clean Energy) program. This is a program whereby the county lends you the money for a renewable energy project and then you pay them back over the life of the project. The county gets the money to do this by issuing a taxable municipal bond†. Chautauqua County has an A1 credit rating from Moody's Credit Rating Service, so it would get a good rate, say 5%. Geothermal systems last 30+ years, so the county can

* Congress originally attached 17 crosscutters to the financial assistance provided by the Clean Water State Revolving Fund under Title VI of the Clean Water Act when Title VI was adopted in 1987. Happily, they let them expire in 1996.

† The bond is taxable because the beneficiary is a private individual rather than a public entity.

issue a 30-year bond. The monthly payment* on a 30-year loan of $60,000 at 5% interest is $325.

Here is the point: If Chautauqua County had enacted the PACE program, the doctor would not have had to wait 5 years to get back his money; he would be ahead in the first month. In Month 1, he would pay $325 but he would save $1,000; so he would have a net gain of $675 in the first month and every month thereafter for the next 30 years!

Here is an even bigger point: Not all of us mortals can write a check for $60,000. Few are willing to wait 5 years to break even. But if I could *spend $325 and save $1,000 in the first month,* I would do it tomorrow. And I will bet quite a few other mortals would do likewise.

The Chautauqua story is not an isolated event. When, you might ask, will the people of this country wholeheartedly embrace solar power or other renewable sources of energy? The answer is: When they are cheaper than conventional power.

This principle, driving down the cost of environmental projects, occurs in many settings, some quite subtle. Here is an example of where this principle not only involves driving down the direct cost of an environmental improvement, but also conserving other funds that would otherwise be lost or wasted.

For example, the general pattern of the SRFs across the country is to make 20-year direct loans at 50% of the market rate for municipal bonds. This means that if your community wants to borrow $1 million from your SRF and the market-rate for high-quality, 20-year municipal bonds (all SRFs have AAA ratings!) is 4%, your SRF will make you a 20-year loan at 2%. Here is how this might happen in at least 34 of the 51 SRF programs.

First, the SRF will issue a $1-million bond for 20 years at 4%. On this bond, they will pay an annual debt service of $73,582. Then they take the $1 million and lend it to your community for 20 years at a rate of 2%, which will require your community to make annual debt service payments of $61,157. The difference, $12,425, is the subsidy. Now, where does the SRF get the money to pay your subsidy? The answer is that it must either keep funds on hand (i.e., not being used to clean the country's water!) in investments, or it must use the subsidized loan repayments from other communities. In either case, there is a cost. This cost can be calculated simply by imagining that the funds are deposited in a 20-year CD at a reputable bank. Let us say that this reputable bank's 20-year CD rate is also 4%. In this case, the SRF would have to invest $310,625 in a CD to pay your subsidy.

Lesson 1: The SRF has just withheld $310,625—more than 31% of the money you borrowed from them—from the goal of cleaning our country's water, just so they could pay your subsidy. (See Chapter 15, "The Curse of Subsidies".)

---

* Unlike home mortgages, municipal bond payments are semi-annual interest and annual principal. So when we refer to "monthly payments" on a municipal bond, we are referring to the sum of the two payments divided by 12.

Now, here is Lesson 2. Recall that the SRF loan with the 20-year term and the subsidized 2% interest rate costs you $61,157. If, instead of borrowing from the SRF, let us say that your community issued a $1-million municipal bond. But, let us say that they issued a 30-year bond at a slightly higher rate of 4.5%*. In this case, your annual debt service payment would be $61,392. In other words, the difference between a market-rate bond with a 30-year term and a direct SRF loan for 20 years with a subsidized rate of 2%—which wastes $310,625—is $235. If your community really needs that $235, the subsidy cost would only be $5,875. So, the principle of driving down the cost of environmental improvements is not always straightforward—but it is consistent.

That is why this principle is so important. If we can drive down the cost of environmental projects or renewable energy projects, more people will undertake them. So, please—as you read the remainder of this book—always remember that our goal is to provide the greatest environmental benefits to the largest number of people at the lowest possible price. And, we achieve the lowest possible price by relentlessly driving down the cost of financing environmental projects.

* Interest rates generally rise as terms extend.

4

Policy Principles of Environmental Finance

One would think, at first blush, that there must be at least 20 or so policy principles that should apply to environmental finance. There may be. As I wrote at the beginning of Chapter 2, my list of principles is a living document. The longer I work in this field, the more principles I can identify. But after almost 25 years, I only see two real policy principles, but they are certainly important. And, furthermore, as I witness various federal, state, and local attempts to deal with environmental problems, I do not see that either of these two principles is readily observed. As I write this at the end of 2013, I have had the occasion to observe the State of Maryland wrestle with the problem of stormwater for the past three years. Regulation is clearly the order of the day. The nine largest counties and the City of Baltimore have all been issued National Pollutant Discharge Elimination System (NPDES) permits specifying specific stormwater limits. But precious little attention has been given to financial incentives and how they would work in tandem with the applicable regulations to actually reduce stormwater run-off.

That said, here are the two policy principles applicable to environmental finance. You will note that there is a twist on Principle 4, the one about the polluter paying.

Principle 3: Regulation and financial incentives are complementary policies. Carrots and sticks work best together. They should be used in tandem. Regulate, yes; but create financial incentives as well, especially for those for whom compliance is an economic hardship or not required.

This is an excellent principle but it needs to be put in context. The specific context is basically "where necessary." Sometimes it is important to pair an environmental mandate with a financial incentive, but sometimes it is just not necessary.

This is a very complicated issue. Probably the simplest example of this principle is the story about Jack and Diane, their wood stove, and their asthmatic baby girl, which appeared in Chapter 1.

If you want to get rid of wood stoves, the first step is to focus on where the problem lies. In this case, it is in *non-attainment areas*—defined areas with inferior air quality, which, in this case, are caused largely by old wood stoves. The second step is to regulate: Ban them in non-attainment areas. The third step is to offer 100% grants to the relatively few people like Jack and Diane who live in a non-attainment area, are poor, have no other source of heat, and have a household member with pulmonary disease. This is the smart way to address the problem: *Both* regulate *and* provide financial incentives where needed. Do not just create a meaningless tax credit and walk away thinking you did something.

Unfortunately, there are not very many other simple examples.

As noted in Chapter 2, the State of Maryland has applied strict regulation to its nine largest counties and the City of Baltimore when it comes to stormwater run-off. But they have not accompanied these regulations with any financial incentives. Moreover, their mandate to these local governments contains the proviso that they must not only create their own local stormwater agencies; but must also fund these new agencies with a local tax or fee.

The stormwater problem is to retard the velocity with which rain reaches the receiving body of water, which, in Maryland's case, is the Chesapeake Bay. You do this by capturing rainwater. As it runs off your roof, you can capture it in a cistern or a retention pond. If you have a large flat roof, you can put soil and plant vegetation on the roof, which will retard the run-off. (It will also insulate your building!) If you have impermeable surfaces such as driveways, sidewalks, and parking lots, you can replace the impermeable surface with a permeable one that will let the rainwater gently percolate through. Changing impermeable surfaces to permeable ones is not cheap. Neither is building retention ponds or cisterns, and neither is building green roofs or rain gardens.

The point here is that the nine counties and the City of Baltimore do not have enough *public land* upon which these stormwater retarding projects can be built. If they are going to comply—and eventually they must—then they must find a way to get *private landowners* to undertake such projects. How are they going to do that? Are they going to regulate, period? Are they going to pass a local ordinance that says, for example, anyone owning property with more than 1 acre of impermeable surface must reduce the impermeable surface by 20%?

How about the owner of a small strip mall with a couple of vacant stores and a 2+-acre parking lot? The owner clearly is not doing too well with this property. So, is the county going to force him to spend his own money to tear up about 10,000 square feet of his parking lot and replace it with a permeable surface? I doubt it.

What if they offer him a financial incentive? Here is where Core Principle 2 comes in: Driving down the cost of environmental projects will result in more such projects getting done. In other words, the larger the financial incentives, the less such private landowners will howl and will get their projects done. Clearly, if the financial incentive is 100% of cost, the landowners have no grounds to complain.

The folks in Maryland have not quite caught on to this principle yet, but they will.

So, there are clearly occasions when regulation should be coupled with financial incentives.

But earlier, we said "when necessary." So, when are financial incentives *not* necessary?

The Clean Air Act (CAA) regulates vehicle emissions as well as emissions from major industrial air polluters such as power plants. There are no financial incentives attached to any of these enforcement efforts. If there were, what would be the point? If the government gave the automakers and the power companies financial incentives, they would undoubtedly use them to pay for the cost of their environmental regulations. That means that the cost of cars and the cost of electricity would not have to rise. But is this a good idea from an environmental perspective? Should we encourage the use of automobiles by artificially depressing their cost? Should we encourage people to use more electricity by subsidizing its true cost? I think that any environmentally conscious soul would answer no to both questions. This example conveniently leads us to Principle 4, the one about "the polluter should pay." Because as you can see from the previous example, it is not the automakers or the power companies that are the polluters; rather, it is us.

But before going on to Principle 4, let us take a look at some vexing regulation/incentive issues that occur under the Clean Water Act (CWA).

As you know, the CWA regulates point-sources of water pollution through the issuance of NPDES permits. If you have an effluent pipe, whether you are a private company or a government, you must have an NPDES permit. If you are a publicly owned (i.e., government) polluter, you get very generous financial incentives from the State Revolving Fund (SRF). If you are a privately owned polluter, you get nothing.* Go figure!

But that is not all. If you are a non-point-source polluter—whether public or private—you are *not* regulated at all: Compliance is voluntary. There is no stick, but you have many financial incentives available to you. There are lots of carrots.

The problem is the word *voluntary.*

Here is the conundrum in a nutshell. According to the United States Environmental Protection Agency (EPA), "In the 2000 *National Water Quality Survey,* states reported that agricultural non-point source (NPS) pollution is the leading source (sic) of water impacts on surveyed rivers and lakes. . . ."† In the highly sensitive Chesapeake Bay, 38% of the nitrogen comes from agricultural run-off.‡

Preventing agricultural NPS pollution is voluntary. You cannot force a farmer to do anything that will lessen the run-off from his farm. That said, the SRF offers farmers low-cost loans to do so. But they are *loans.* Farmers have to pay them back out of their own pockets. They are not like sewer agencies that just raise their rates. Farmers cannot raise the price they get

---

* With one very limited exception that, itself, has been used only once since the SRF (State Revolving Fund) was created in 1987.

† "Protecting Water Quality from Agricultural Runoff," USEPA. EPA 841-F-05-001, at 1.

‡ Chesapeake Bay Program, Phase 4.3 Watershed Model, 5/11.

for the crops they grow or the livestock they raise. Other than just a wealthy few who might undertake projects from the goodness of their hearts, or a few more who can somehow use a pollution abatement project to increase production, why would any farmer in his/her right mind *borrow* money to prevent pollution? This problem is similar to the stormwater issue. With the stormwater issue, you had (potential) regulation coupled with (potential) financial incentives. The notion there was that you can probably compel private landowners to undertake stormwater remediation projects if you give them enough financial incentives. However, on the other hand, here, in the NPS area, we have no regulation at all. We have no stick.

As time rolls on and the percentage of water pollution from agricultural run-off continues to increase, the regulation of farm practices will gradually start. There will eventually be a stick. When this happens—as is the case with stormwater—the success of such programs will depend on the size of their financial incentives.

And now on to Principle 4.

Principle 4: Forget about divisive canards like "the polluter pays." We are all polluters, both directly and indirectly. We are all going to have to pay. And the more we pollute, the more we should pay.

There is not really much that needs saying about this principle. It is pretty self-explanatory. You saw, a few lines above, the examples about automobiles and electricity. We drive the cars. We pollute, directly in this case. We use electricity. Most electricity is generated by the burning of fossil fuels. So, when we use electricity, we pollute, albeit indirectly in this case.

Now that we have established the fact that we are all polluters, let us look at the last part of this principle: That the more we pollute, the more we should pay. Later on in the book we will discuss ways to raise money for environmental finance programs. But for now, let us take a look at an example of where the more one pollutes, the more that person will pay.

Consider the imposition of a tax on vehicle emissions. Vehicle emissions speak directly to the issue of the air deposition of nitrogen causing water pollution. They also speak to the production of CO_2 (carbon dioxide). Using Maryland again as an example, its automobiles produce 14.885 tons of CO_2 each year. Its trucks produce another 14.824 million tons of CO_2.

Let us posit a flat vehicle emissions tax of a \$40 for trucks and a sliding scale tax of up to \$40 based on the average of each vehicle's Air Pollution Score and Greenhouse Gas Score.

The Air Pollution Score and the Greenhouse Gas Score are indices developed by the EPA to rank air emissions by vehicle model. Both scores run from 1 to 10, with 10 indicating vehicles with the lowest emissions.

Here is how such a tax could work: First, we would take both scores and then average them. Then we would subtract that number from 10, multiply by 0.1, and then multiply that number by the maximum surcharge to determine the specific tax on any particular vehicle. For example, the Toyota Prius has an Air Pollution Score of 8 and a Greenhouse Gas Score of 10. The average is 9; 10 minus 9 equals 1; 1 times 0.1 is 0.1. Finally, 0.1 times the $40 tax rate is $4. So, an owner of a Prius would pay a $4 tax per year.

All in all, the median average of the scores should be 5, which means that the average personal vehicle owner would wind up paying a $20 emissions excise tax each year.

With its 2,583,261 automobiles, at an average vehicle emission surcharge of $20 per year, Maryland could raise another $52 million. This would support annual debt service payments on some $800 million of bonds for environmental projects.

With its 1,849,201 trucks, at $40 per year in vehicle emission surcharges, Maryland could raise about $74 million. This would be enough to support the annual debt service payments on some $1.138 billion of bonds, the proceeds from which would be used to fund stormwater grant programs in counties throughout Maryland.

So, this vehicle emissions surcharge, on both cars and trucks, could generate a total of about $126 million, which would pay the annual debt service on some $1.938 billion of bonds, which is over 36% of the total that Maryland needs to fund its environmental finance programs.

A final word about "the polluter paying": When the US Congress amended the CWA in 1987, they replaced the construction grant program, which they had created in 1972, with the Clean Water State Revolving Fund (CWSRF). Now the construction grant program was limited to Publicly Owned Treatment Works (POTWs)—no private point-source polluters. So, the CWSRF was made to apply the same way: Only POTWs were eligible for financial assistance—no private point-source polluters. What was going on here? Is this just a case of following suit as close to the construction grant program as possible? Or was their some *animus extant* in Congress against private polluters?

My recollection is that it was both. Many wanted the CWSRF to look as much as possible like the construction grant program. Certainly understandable. But there were definitely some who felt that private industry should be made to pay for the pollution they created, without any type of financial assistance from any government. There is always some clique of anti-business representatives in any government. In this case, wittingly or not, they prevailed.

The final irony of this "polluter pays" mind-set cropped up in the passage of the American Recovery and Reinvestment Act (ARRA) of 2009. This act provided a windfall of new money for the CWSRF ... but there was a string attached: The CWSRF borrowers were required to "Buy American." So here is the wonderful irony of the US Congress telling CWSRF borrowers that they must patronize the same US industries whom the Congress had shut out of the CWSRF. "We want you to buy from them, but we won't even think of offering them help paying for their water pollution compliance measures." Clearly, the Congress understands the nexus between purchasing from US companies and the jobs that creates. What is lost on them is the jobs that are lost because companies cannot afford to pay for their NPDES permit requirements.

Let me give you a quick example: Let us say that a company has to undertake a $10-million project to comply with its NPDES permit. Now, most companies cannot borrow for long terms unless they have a shiny piece of real estate to offer as collateral. Here, the collateral would be (used) water pollution control equipment. Collateral value: zero!

So, let us say that all the company can get is a 5-year loan at 9% interest. As such, their annual payment would be $2,570,920. If, on the other hand, they could borrow from the SRF—even at an *un*subsidized rate of 4%—for 20 years, their annual payment would be: $735,818. The SRF payment is *71% lower* than the conventional loan.

A lot of jobs might have been saved if the polluter had been able to get a break on his payments. He certainly would have paid. Paid in full. But he definitely could have used a break.

5

Management Principles of Environmental Finance

This is a continuation of the 23 environmental finance principles set forth in Chapter 2. These are the *management principles* that apply directly to environmental finance. There are six of them. Most of them—like most good management principles—are really just common sense. You should not be surprised seeing them.

And, also like most good management principles, they are quite simple. No complexity here. These principles are very straightforward.

That said, many are fiendishly difficult—if not impossible—to implement.

As the first two address the same issue, we will look at them together. Here they are:

Principle 5: Put one person in charge. Make that person accountable. Be prepared to fire.

Principle 6: Advisory committees, yes; executive committees, no.

These are two of the principles that are simple, common sense, and probably impossible to implement.

All governments have departments. In most, the "environmental" department is the regulator. There are some exceptions to this rule. Both the US Environmental Protection Agency (EPA) itself and the Maryland Department of the Environment house not only the regulators, but also those who provide the financial incentives. On the other hand, most states—Maine, New York, Texas, Ohio, Pennsylvania, and Virginia, to name a few—separate the financial incentive function from the regulatory function. So, immediately you know that in these latter states, there are not two people in charge; there is no one in charge. There is no single person to whom the governor can turn and ask if his/her environmental programs are working to their maximum benefit.

Going back to the notion that all governments have departments, we need also to say that "the environment" is not the exclusive province of any one of them. In the federal government, we have the EPA, of course, but then the Department of Energy, the Department of Agriculture, the Department of the Interior, and the Department of Transportation. Then there are also little chunks of responsibility in the Departments of Commerce and Housing and Urban Development; and there are the lesser luminaries who also have a seat at the table—at least sometimes—like the National Oceanic and Atmospheric

Administration (NOAA) and the Bureau of Ocean Energy Management (BOEM). So, who is "in charge" of "the environment"? The answer: no one.

Of course, it can always be said that the President is in charge. But, of course, he is not really in charge. He is responsible for the environment but he is certainly not in charge—and definitely not on a day-to-day basis.

Does it make sense to put one single person in charge of defense? Yes. Transportation? Yes. The nation's lands (Interior)? Yes. The environment? Yes, of course, but it is never going to happen. (Oh yes, there is a head of the EPA, but is that person in charge of the environment of the United States? No.)

This is not just a federal problem. It exists at the state and local government levels as well. At the state level, you have the environmental regulators, then the natural resources people, then agriculture, then planning, and maybe more. At the local level, you have the environmental office, the planners, and the parks people.

In short, at every level of government, there is no one person in charge of the environment. And because of our inherent—and long-standing—government structures, there is unlikely ever to be.

And so, if no one is in charge, what about an "executive committee" composed of the heads of all these agencies?

I am only going to make one point here: That as long as the members of a committee have statutorily differing legal responsibilities, the words "executive committee" are a contradiction in terms.

The closest you can come to real executive authority is when the President, or a governor, or a county executive convenes an "environmental cabinet," if you will, and takes personal responsibility for its actions. But as noted before, this is not going to take place on a day-to-day basis.

So, now, what about "advisory committees"? Advisory committees can, I guess, have many definitions. But to me, a real advisory committee is one composed of both experts and stakeholders. The experts should be able to advise on the best and most effective way to get things done. The stakeholders should be able to advise on (and warn of) the impact that government environmental decisions will have on the lives and property of the people. With that understanding, and in this limited sense, advisory committees can provide a great service to us all.

Principle 7: Put everything on the web. Make strategies and programs fully transparent.

As Francis Bacon said, "Knowledge is power." The more people know about what is being done to protect their environment, the more empowered they will feel and—unless they are just anti-environment—the more they will support those efforts.

Many years ago, as a young man, as I was preparing the documents for the first bond issue of the state agency I ran, I sought the advice of a truly great man, the New York State Budget Director, Peter C. Goldmark, Jr., who said,

"Always err on the side of disclosure." Ever since then, the concept of full disclosure has permeated my career in finance. I could not even conceive of taking any action that affected the public without full and timely disclosure of all relevant information. And, when I was not sure whether or not the information was truly relevant, I would follow Goldmark's advice and "err on the side of disclosure."

That conversation with Goldmark was many years ago. And now it never ceases to amaze me to pick up a financial news report today and find the Securities and Exchange Commission (SEC), or the Financial Industry Regulatory Authority (FINRA), or the courts still chasing down government and corporate officials who have failed to disclose some seriously material facts in an important financial transaction.

So, do not take transparency for granted. Check into it and insist upon it.

Principle 8: Be prepared to change. If one strategy does not work, trash it and adopt another.

Some people become wedded to programs and policies. Without continually—or at least periodically—scrutinizing the success of such, this is a bad idea.

Since the Clean Water State Revolving Fund (CWSRF) was created in 1987, most states have funded their non-point-source projects through programs called *linked deposits* or *loans-to-lenders*. In programs like these, a farmer might apply to his CWSRF for a $100,000 loan to build an Animal Waste Management System (AWMS) on his farm. In such a case, he would be directed to a list of qualified banks. He would then negotiate an interest rate and term with the bank, and the bank would make him the loan. The role of the SRF (State Revolving Fund) in this program is as follows: The SRF will deposit the same amount as the loan ($100,000) in the bank in a Certificate of Deposit for the same term as that of the loan that bears an interest rate of, say, 2% below market. The understanding here is that the interest rate that the bank charges the farmer is also 2% below market. (Of course, there is no real way of knowing that.) It is also important to note that the bank takes the risk of loss. So, if the farmer does not pay, the bank loses its money.

There are two problems with this program. The first is obviously the problematic interest rate charged the farmer. The second, and more important, problem is the term. Banks do not to like to make loans like this for terms of more than 5 or, maximum, 7 years. Now the asset, or service, life of an AWMS is at least 20 years. Governments are used to financing assets not at the whim of the bankers, but over the full service life of that asset. If they are financing a school bus that will last 10 years, the school board finances it for 10 years. Not 5 years. And certainly not 20 years.

So the SRF should be in the business of financing things like AWMSs over their full service life. In this case, if the bank made a 4% loan to the farmer for 5 years, the farmer would have to pay $22,462 a year. On the other hand, if the SRF had loaned the money at 4% for 20 years, the farmer's annual payment would only be $7,358. Please remember the axiom of environmental finance that the lower the cost of a project, the more such projects will be undertaken.

Suffice it to say that these loans-to-lenders programs should have been trashed years ago and replaced by direct long-term loans from the SRF. Is it any wonder that of the $100+ billion of financial assistance provided by all the SRFs, that less than 4% has gone to non-point-source projects like our farmer's AWMS?

So, the moral of the story is to periodically evaluate environmental finance programs, and, if they are not getting the job done, trash them and replace them with ones that really work.

Principle 9: Develop new strategies. Issues like non-point-source water pollution and small-source air pollution cannot be tackled with today's legal and administrative tools. New ones must be designed and adopted.

As you read in Chapter 1, between the adoption of all the major environmental laws 30 to 40 years ago and now, we have picked a lot of the low-hanging fruit. We have just about rid our air of lead and drastically reduced vehicle emissions. And we have restored our major waterways to health by vigorously attacking (and heavily funding) the problem of municipal sewage.

We are now looking at the second wave of problems. These are the literally hundreds and thousands of small non-point sources of water pollution and small sources of air pollution.

A few paragraphs above, I suggested that loans-to-lenders programs should be trashed in favor of direct loans from SRFs. I need to admit that this suggestion was more than a little fatuous.

SRFs are not equipped to lend to hundreds of thousands of farmers and other non-point-source polluters. SRFs are used in making multimillion-dollar loans to Publicly Owned Treatment Works (POTWs) whose creditworthiness they can check out online with a subscription to one of the major credit rating agencies. They do not have the staff to deal with thousands of farmers. And they have no clue as to how to evaluate a particular farm loan credit. What if it is a personal loan to a farmer with a middling credit score? What if it is a loan to a farmer with an execrable credit score, but the project is collateralized by the farm itself? No SRF staff have ever had to deal with matters like these.

As you will see as you continue to read this book, there are certainly effective strategies for dealing with these issues. The problem is that we must now start designing and adopting these new strategies.

Principle 10: Do it now. Everything costs more tomorrow. The impact of inflation.

This principle is another simple, commonsense matter. Everyone intuitively understands the time value of money.

I always tell the story of buying a quart of milk for 23¢ when I was a little boy, versus what I pay for it today. Or when my sister bought a great three-bedroom house for $35,000.

Everyone has stories like these. This is inflation in the real world. Things will always cost more tomorrow than they do today.

Now, as we all know, the cost of everything rises over time, and we must ensure that projects that need doing today get done today. That they are not put off until tomorrow, when they will be even more expensive.

As I said, a simple, commonsense matter. We just need to be on guard that it actually happens.

6

Revenue Raising Principles for Environmental Finance

Below are five principles of raising funds that apply to the financing of environmental projects.

When it comes to raising money for environmental projects, *size* and *stability* are the two watchwords. The reason for this is that one of the optimum uses for "raised funds," if you will, is to pay the annual debt service on bonds issued to finance environmental improvement projects. Almost all public debt is level payment debt. This means that the annual payment is the same every year. That is one reason why the "raised funds" should be stable. This is not just a practical matter, rather it has a huge impact on the rate of interest you will pay on such debt.

Virtually all public debt in the United States is rated by one of the three major international credit rating agencies. Before they issue a rating on a bond issue, they investigate the source of repayment of those bonds. The larger the revenue stream and the more stable it has performed over time, the higher the rating the bonds will receive that are supported by such a revenue stream.

"Large" has other—obvious—benefits as well. If you finance an environmental project with, say, a 30-year bond at an interest rate of 4%, your annual payment will be about 5.7% of the principle amount. So, if your environmental tax/fee generates $57 a year, you can pay for a $1,000 project. If it generates $57,000, you can afford a $1-million project. And, if it generates $57 million, you can afford a $1-billion project. So, the size of the revenue stream directly affects the size of the project you can finance, and both the *size* and the *stability* of the revenue stream affect the rate of interest you will pay on the debt. Please keep these notions in mind as we go through these principles of raising money for environmental projects.

Principle 11: Raise money from many small charges, fees, or taxes—not one big one. Many small sources of money are more stable than one large one.

This principle is generally true, but not always. I think you can all see that you can certainly raise the gasoline tax. The problem occurs when the price of gasoline spikes—maybe up to European levels of $7+ per gallon. In such a case, people do not drive as much. Gasoline sales plummet. And so do the gas tax revenues.

The same is even true of a sales tax. You can pledge part of a sales tax revenue stream to the repayment of an environmental bond. But what happens when a recession hits? General sales plummet. And so do sales tax revenues.

It would be better to create several smaller revenue streams. Maybe a vehicle emissions tax. Also a boat registration surcharge. And a small bridge/tunnel toll increase. Also a tax on marine fuel. In other words, a series of revenue streams can stabilize the income used to pay for debt service. All these independent revenue streams will not be dependent on one economic event; thus they will be more stable.

The exception to this rule is when the tax is on an essential service. Maryland's *flush tax* is a good example. Anyone in Maryland whose building is connected to a sewer main or a septic tank must pay. This means that ALL building owners must pay—unless, of course, they want to disconnect their toilets from the sewer or the septic system. This is such an unpleasant alternative that even the rating agencies would give full credence to a bond backed by the *flush tax.*

Principle 12: Once collected, put all the environmental money in one basket. Do not fragment it or piddle it away.

In 2006, I had occasion to review the budget of the Maryland Department of Natural Resources. To my astonishment, that department collected about two dozen small fees. Normally this would be a good thing. You recall that many small revenue streams are often more stable than one large one. But in this case, each of these small funds could only be spent in a certain way. The duck fund could only be spent on ducks. The oyster fund could only be spent on oysters. The rockfish fee could only be spent on rockfish. And so on.

The problem here is what happens if there is an overabundance of oysters—so we do not have to spend the oyster fund trying to replenish them—while there are few rockfish that desperately need replenishing? There is a legal firewall between each of these little funds. This is bad finance policy. I realize that the oystermen fought hard for the oyster fund and that the hunters fought hard for the duck fund; but there should be an amicable process whereby the funds could be directed each year to where they are most needed and then later replenished.

Principle 13: Change behavior while raising money. Do not tax all equally; tax the polluters more while rewarding energy efficiency and green practices.

In Chapter 4 we discussed a vehicle emissions tax based on two US EPA-based matrices of emissions and fuel consumption. The tax was a sliding scale from $4 to $40. The most environmentally friendly cars paid $4, while the least friendly paid $40. I think you see the principle here. I also think you see that the difference between a $4 fee and a $40 fee is not really going to change many people's behavior. People are not going to flock to low-emission, fuel-efficient cars just to save $36. But on the other hand, if the sliding scale tax went from $4 to $4,000, it might turn a few heads—especially if it were an *annual* fee.

The human race needs to break its dependence on automobiles... as unpleasant a thought as this is. Thus, raising gasoline taxes is always a step in the right direction. This is, as you can see, an extension of the "polluter pays" principle. But in this case, it is justifiable.

The same thing is true with respect to climate change and greenhouse gases. The human race pumps about 44 billion tons of carbon dioxide (CO_2) into the atmosphere. This gas—like other greenhouse gases—traps the sun's rays in the atmosphere, amplifying their effect. The planet is still on the rebound from the last ice age about 20,000 years ago. The 44 billion tons of CO_2 are not helping.

Governments need to require people to reduce the amount of carbon they emit. One of the most effective ways of doing this is to tax carbon emissions. For the past three decades, governments around the world have experimented with cap-and-trade programs where emissions are simply reduced by regulation, allowing polluters who have an inexpensive means of reducing emissions to sell some of their emission reduction credits to those whose reductions are more costly. This system has met with mixed results. Sooner or later, however, governments will probably come home to taxing carbon emissions. This will become especially true when they realize how valuable these funds can be in stimulating the growth of reliable renewable energy technologies.

So, if you want people to emit less carbon, tax those emissions.

Principle 14: Use "dedicated revenue streams" (such as annual taxes or fees) to finance capital, not operational, expenses.

As noted above, one of the best uses of a large, diverse revenue stream is to pay off environmental project debt.

Much has been said about the State of Maryland's infamous Bay Restoration Fee, or flush tax. Maryland uses these funds to provide 100% Enhanced Nutrient Removal (ENR) grants to its 66 wastewater treatment plants that account for 95% of all the nitrogen entering the Bay from sewage. The flush tax now takes in about $120 million a year. Certainly one of the great genius aspects of the flush tax is that the funds are not used as direct grants. Maryland has about a $1.3-billion ENR problem. So, at $120 million a year, it would take them more than 10 years to deal with it. As it is now, the ENR problem will be solved by 2017. This is because the flush tax legislation provides that the $120 million be pledged to bonds that are issued to fund the ENR projects. So, the ENR projects can get done today, and the projects can be paid for over time by the annual receipts of the flush tax.

This is an extremely efficient means of paying for major environmental programs like the $1.3 billion of ENR projects.

Principle 15: Make it as painless as possible.

As noted above, at any level of government, raising the rates on a general tax—like the income tax, the property tax, or the sales tax—will set up howls of protest across the country. Virtually every newspaper will editorialize against it. It will also mobilize armies of lobbyists in Washington, DC, or state capitals, who will roam the halls of Congress and state legislatures persuading members to oppose it.

Here is another interesting little lesson from Maryland. In 2007, the State created an entity called the Chesapeake and Atlantic Coastal Bays Trust Fund, which most people in Maryland call the *green fund*. A large part of the revenue that finances the green fund comes from a politically ingenious tax: on rental cars!

Of all the people who rent cars in Maryland, how many of them do you think are Marylanders? Not too many. So, did the Maryland newspapers oppose it? No. Did armies of lobbyists descend on the Maryland General Assembly? No. Oh, sure, the car rental companies opposed it, but that is small potatoes in terms of opposition. Hertz, Avis, et al., are not known for their high-powered lobbyists.

So, most car renters come from out of state. Did the out-of-state newspapers editorialize against the car rental tax? No. Would any member of the Maryland General Assembly have cared if they did? No. So the tax passed because it was painless.

There is another source of revenue that is also painless—or almost painless. That is the revenue that derives from a tax-increment district (TID). TIDs are most commonly found in brownfields remediation projects. Think of a smelly old urban landfill. Imagine what the property values are on the parcels surrounding the landfill for several blocks. These property values are typically greatly depressed.

So, the local government decides that it is going to get rid of the landfill by covering it with impermeable material and then covering that with several feet of topsoil. As a matter of fact, the government decides to turn the old landfill into a golf course. Now, once the smells and the eyesores from the landfill are gone and there is a pristine new golf course there, the property values will rebound dramatically. Let us say that a small pre-golf-course parcel was worth $10,000. After the golf course is built, this same parcel is worth $50,000. Let us now say that the pre-golf-course real property tax rate was 1%. So, before the golf course was built, the local government got $100 in tax revenue from this little parcel. Now, however, it is worth $50,000. The real property tax rate is still the same, but now the local government gets $500 instead of $100.

Now, let us say that the local government formed a TID as the financial means of implementing the landfill-to-golf-course project. As such, they would have created the district such that it encompassed not only the old landfill, but also the several blocks surrounding the landfill on all sides where the real property values were greatly depressed.

The legal charter creating the TID would have three critical elements: (1) the TID could issue municipal bonds to finance the project, (2) the TID would receive the *increment* in the local property tax revenues, and (3) those incremental tax revenues would be used to pay the annual debt service payments on the TID's project bonds until the project bonds were retired, at which time the revenues would revert to the city.

So, the TID would get the incremental $400 of property tax revenue on the little parcel, as well as all the other incremental property tax revenues on all the other properties in the district. The owners of all the property in the TID now have some seriously valuable real estate. True, they have to pay more in taxes (not because the rates went up, but rather because their assessed value went up). But they should feel somewhat compensated for this higher tax payment because of the increased value of their property. That is why we say that TID financing is *almost* painless.

Here is a real-world example of how a novel type of TID might work in a place like Maryland to help them pay for the Bay:

Properties along the shoreline of the Chesapeake Bay and its tributaries are among the most coveted in the state. As a result, they are extremely valuable assets to their owners. And the principal reason for this value is the Bay itself. As such, owners of shoreline property have a strong interest in the health of the Bay. If the entire Bay were nothing more than an open sewer, their property values would plummet. Therefore, because these property owners receive direct economic benefit from the Bay—far more than the average citizen of Maryland, it is only fair to ask them to shoulder a slightly larger share of the cost of maintaining the health of the Bay and its tributaries.

According to the Maryland Geological Survey, there are 6,776 miles of Chesapeake Bay (and tributaries) shoreline in Maryland. The shoreline can best be measured by the Critical Area, which is the first 1,000 feet in from the shoreline; 1,000 feet constitutes 0.189 miles. So that means there are 1,283 square miles in the Critical Area in Maryland, which, at 640 acres to the square mile, means that there are 821,333 acres in the Critical Area in Maryland.

Using an estimated assessed value of only $100,000 per acre, means that the current assessed value for the entire Critical Area is $82.13 billion. Given the attractive character of such property, the value number is likely to be closer to $200,000 per acre.

The state portion of the real property tax rate in Maryland is 0.00112. In most areas in Maryland, the combined county and local real property tax is about $1 per $100 of assessed valuation, or 1% of the assessed value. As such, the state real property tax is only about 10% of the local tax. At an assessed value of only $100,000 per acre, the State takes in about $91,989,333 in real property taxes from the Critical Area.

So, let us say that the State of Maryland creates a "Critical Area TID" out of all 821,333 acres!

Now after the project was finished in our brownfields example above, all the properties in the TID were reassessed to reflect their higher value because of their proximity to the golf course versus the landfill.

But in our Critical Area TID, there is no project. There is no event. So, there is no reassessment. Where does the money come from? Natural events.

Homes in the United States change ownership approximately once every seven years. During a seven-year period, let us say that property values in the Critical Area increase by 20%. With normal housing turnover, that means that in seven years, the Critical Area TID should be taking in about $18 million a year. These revenues could then be pledged to pay annual debt service payments on bonds issued to fund Bay restoration projects.

In this case, such a TID would be much less painful than the brownfields approach. There, existing landowners got hit with higher tax bills. In the Critical Area TID, only the new owners would pay the higher tax, not the existing owners. So, much less of a problem.

There is another interesting option available to the State with the concept of this Critical Area TID. The State could make the TID both a TID and a Special Tax District (STD). An STD is a very simple concept: It is simply a discrete geographical area where the property owners pay a certain tax that property owners outside the district do not have to pay.

Here is how this might work: As part of its greater efforts to help finance Bay restoration, the Maryland General Assembly could also create a Chesapeake Critical Area Special Tax District comprising the same 821,333 acres of the Critical Area as are in the TID. It could then impose a Bay Restoration Surcharge equal to the current state tax. This would generate another $91+ million, all of which would go to Bay restoration.

How would this work? On a Bay-front home assessed at only $500,000, a county and local combined property tax of 1% (0.9% for the county, 0.1% for the state) would equal $5,000. The State portion of the tax would equal $500. So, through the mechanism of the STD, the State could double its share of the tax to 0.2% in the form of a Bay Restoration Surcharge. The additional $500 would, of course, go to the STD, which, in turn, means that it can be used to support debt for Bay restoration projects, just as in the case of the TID.

An additional $500 Bay Restoration Surcharge would bring the total property tax bill to $5,500, of which the new Bay surcharge accounts for only 9.1%. So, such a surcharge—on the properties that economically benefit the most from the Bay—would only amount to a 9.1% tax increase.

Creating a Special Tax District out of the Critical Area in addition to creating a Critical Area TID is certainly not painless. It would definitely generate heated opposition from wealthy landowners in the Critical Area. But it *is* an option.

7

Sources of Revenue for Environmental Finance Programs

The previous few chapters discussed some principles to be observed when raising money for environmental finance programs. In doing so, several important examples were cited. In Chapter 4, we described a *vehicle emissions tax* based on two matrices of emissions and of fuel consumption developed by the US Environmental Protection Agency (EPA). In Chapter 5 we discussed several more. Among them were Maryland's infamous *flush tax,* and, more importantly, the *Chesapeake Bay Critical Area Tax Increment District* and its more aggressive counterpart, the *Chesapeake Bay Critical Area Special Tax District.*

In this chapter we are going to present some more possible sources of revenue for environmental finance programs. Some, like the Air Passenger Mileage Tax/Air Freight Tax, are very large. Some, like the *Fishing License Stamp,* are quite small. And, although the *Air Passenger Mileage Tax/Air Freight Tax* and the *Fishing License Stamp* are reasonably painless (both because they are quite small, and the latter because its "payors," the fishermen, are usually sympathetic to water quality initiatives), most of the proposals below actually violate the principle of "making it as painless as possible" because some of the time it is just not possible to make money raising painless.

So, here are some ways to raise money.

Air Passenger Mileage Tax/Air Freight Tax

The Bureau of Transportation Statistics indicates that in the 2010 flight year (May 2010 to April 2011), there were 817,743,551,000 air passenger miles flown in the United States, including both domestic and international flights.* If passengers were charged 1¢ for each mile they flew, that would generate more than $8 billion each year. If these funds were allocated on a per-capita basis, California, our most populous state with 11.91% of the population,† would get about $974 million a year. Texas, our second most populous state with 8.04% of the population, would get over $657 million. Even Wyoming,

* US Bureau of Transportation Statistics, US Air Carrier Traffic Statistics through April 2011, available at: http://www.bts.gov/xml/air_traffic/src/index.xml.

† According to the 2010 Census.

our least populous state, with only 0.18% of the population, would get almost $15 million a year.

The annual debt service on a 30-year bond with an interest rate of 5% is about 6.5% of the bond principal. So, if a community issued a $1-million bond with such terms, it would pay $65,000 per year in annual debt service payments. This means that if California used all $974 million of its Air Passenger Tax to pay the annual debt service on bonds issued to support its environmental programs, it could issue almost $15 billion of bonds. Texas could support over $10 billion of environmental improvement bonds, and Wyoming could support about $230 million of environmental bonds.

A final word about the painlessness of this tax: The air travel distance between New York and Boston is 190 miles. Thus, our Air Passenger Mileage Tax would cost a traveler from New York to Boston an additional $1.90. Is an additional $1.90 going to push anyone under the poverty line? No. The distance between New York and Los Angeles is 2,448 miles. Thus, our Air Passenger Mileage Tax would cost a traveler on this route $24.48. Now, the cheapest plane ticket from JFK in New York to Los Angeles costs $1,033. So, that means our $24.48 increases the cost of the ticket by a little less than 2.4%—less than it costs to check a bag!

Now let us look at an Air Freight Tax. In 2011, there were 7,304,348,000 ton/miles of freight hauled by airplanes. If this freight were surcharged $1 per ton/mile, that would generate in excess of $7 billion. California would get about $870 million with which they could support over $13 billion worth of environmental program bonds. Texas would get $587 million with which they could support over $9 billion of bonds. Wyoming would get over $13 million with which it could support over $202 million of environmental improvement bonds.

As you can see, these are big numbers—and they are relatively painless.

Increases on All Toll Roads

This is another Maryland example. In June 2005, the Chesapeake Bay Foundation (CBF) conducted a series of polls among registered voters in Maryland that revealed that 78% of the respondents would support a small tax or fee increase if the money went to Bay restoration. Specifically, 75% of the respondents were willing to pay an extra $10 per year, 66% would pay $25 per year, and 53% would pay an additional $50 to support agricultural pollution abatement programs. This is a solid base of public support that can be used to justify small, incremental increases in taxes and fees, such as modest increases in tolls.

Maryland has six toll facilities. Their names and respective approximate vehicle traffic counts in 2011 were as follows:

Facility	Traffic Count
Bay Bridge[a]	13,000,000
Fort McHenry Tunnel	43,000,000
Harbor Tunnel	24,000,000
Hatem Bridge[a]	4,650,000
I-95[a]	14,500,000
Key Bridge	11,000,000
Nice Bridge[a]	3,000,000
Total	**113,150,000**

[a] These facilities only charge tolls in one direction, so their total traffic count has been divided by 2.

As you can see, a 10¢ additional toll would produce $11,315,000; a 25¢ increase would generate $28,290,000; a 50¢ increase would produce $56,580,000; and a $1 increase would, of course, produce $113,150,000.

It should not go without notice that this $113,150,000 of additional revenue could support over $1,740,000,000 of bonds for environmental improvement projects.

Special Boat License Tag

This is a concept to create a distinguishing environmental tag for boat licenses (like special vehicle license plates, which almost all states have) with a charge of $5 per year. This program could be either voluntary or mandatory.

Several states across the country offer boaters the option of making voluntary donations as part of the boat registration process. The State of Washington, in particular, has an application check-off option that sends funds to a nonprofit organization.

In Maryland, the annual boat registration process, which includes both registration and title, costs $26 per boat. There are approximately 188,623 boats registered in the State of Maryland. If the program were voluntary with a $5 check-off and a participation rate of 10%, the program would generate approximately $94,000 annually. If such a program were mandatory, it would bring in $943,115, which could support environmental improvement bonds of over $14,500,000.

Fishing Stamp

This concept is similar to the special boat tag. It could be either voluntary or mandatory. In this case, you would create a distinguishing fishing license stamp with, again, a charge of $5 per year. In Maryland, several years ago, they issued about 150,000 fishing licenses of one kind or another. As a voluntary program, with a 10% uptake, this program would generate $75,000 per year. If it were mandatory, it would yield $750,000, which could support about $11,500,000 of environmental improvement bonds.

The fishing stamp and the special boat tag are not big money-raising programs but they do serve a useful purpose: They make people feel that they are part of the team, part of the environmental improvement process. They are sort of the opposite of the "change behavior" revenue options. These are actually "affirm behavior" options.

Impervious Surface Fees

We are going to use Maryland again as our example for several reasons. First, impervious surface fees speak to the problem of stormwater. Maryland, and its Chesapeake Bay, have major stormwater problems. Second, Maryland is quite far up the curve in dealing with their stormwater issues. And third, the way they are dealing with these issues has prompted a firestorm of controversy.

Impervious surface fees had their genesis in the 1987 amendment to the Clean Water Act, which was originally adopted in 1972. The amendment required the implementation of a national program to regulate non-agricultural stormwater run-off. A series of scientific studies confirmed that stormwater run-off was in direct proportion to the percentage of impervious surface in the drainage area. The concept, therefore, began to take hold that stormwater control programs could be paid for by those who create the problem, that is, the owners of property with a significant proportion of impervious surface.

The regulation of stormwater run-off takes a variety of forms across the country. In most areas, cities or counties manage the problem directly. In other areas, special stormwater management districts have been created with the legal authority to levy fees to support their activities. When I started researching this issue in 2009, there were approximately 500 communities around the country that implemented stormwater management programs and funded them with impervious surface fees. By now, there are probably over 1,000. These programs are widely dispersed throughout the country but are especially prevalent in certain states, such as Oregon, Washington, North Carolina, Georgia, and California.

In most of these communities, the political leadership initiated the program by empanelling a citizens' committee to formulate a revenue structure for stormwater management. Typically, the citizens' group hired firms to do aerial surveys of their jurisdiction or otherwise calculate the amount of impervious surfaces. This includes roads, roofs, parking lots, driveways, etc.—any surface where rainwater cannot seep into the ground.

Once they knew the amount of impervious surface acreage and the ratio of impervious surface acreage to total acreage, these panels then put this information together with the stormwater program's operating budget and came up with an annual fee based on the amount of impervious acreage.

In doing so, for ease of administration, they estimated the average single-family plot size along with the average impervious surface on the typical single-family dwelling site to arrive at an average impervious surface tax for a single-family home. In most jurisdictions, this number became the centerpiece of the revenue program. In other words, the fee on the typical single-family home became the benchmark for such taxes on all other kinds of properties. Most communities developed specific terminology, such as Equivalent Residential Unit (ERU), or Equivalent Service Unit (ESU), to quantify this level of impermeability. All tax rates for other types of properties were thereinafter measured in ERUs.

For example, Takoma Park, Maryland, imposed an impervious surface fee in 1998 that they call a *stormwater fee*. In doing so, they defined their ERU as 1,228 square feet. Every few years, the City Council resets the fee *per ERU*. Thus, in 2007, Takoma Park's stormwater fee was $4 per ERU per month, or $48 per year. This is, essentially, the base rate. All other property types base their fee on this ratio of $48/1,228 square feet. Thus, for example, the owner of a commercial property with 4,000 square feet of impervious surface would pay 4,000/1,228 × $48, or $156.35, because the impervious surface is 3.26 times larger than that of the ERU.

I took a representative sample of ERUs in communities around the country several years ago. Their respective ERU sizes and annual fees per ERU were as follows:

City/County	ERU Size (sq. ft.)	Annual Fee per ERU
Bend, OR	3,800	$48
Takoma Park, MD	1,228	$48
Raleigh, NC	2,260	$48
Durham, NC	<2,000	$26.04
	>2,000	$54
Chapel Hill, NC	2,000	$39
Orem, UT	2,700	$54
Athens-Clarke County, GA	2,628	$42
West Des Moines, IA	4,000	$33
Lawrence, KS	1,800	$48
	<3,000	$60
	<4,800	$86.40
	>4,800	$120
Reading, MA	2,552	$39.84
New Albany, IN	2,500	$38.04
Milwaukee, WI	1,610	$32

Most communities seem to follow the Takoma Park model. Notable among them are the cities of Milwaukee, Wilmington, Raleigh, Durham, and Chapel Hill. Durham actually has a slightly different system. They use 2,000 square feet as their ERU and charge $26.04 per year for any impervious surface *up to* 2,000. For any such surface greater than 2,000 square feet, they charge a flat fee of $54.

Other communities and one state, Vermont, have taken a slightly different tack. Instead of placing a fee on all impervious surfaces, they just decided to charge for *new* installations of impervious surfaces. Anne Arundel County in Maryland has had a series of such charges. They impose a flat fee, called a *fee in lieu of open space,* for new residential construction of less than 2,000 square feet. A second fee of $1.20 per square foot is charged if the impervious surface exceeds 25% of the total square footage of the individual parcel. Finally, they charge $1.80 times "two times the square footage" for new impervious surfaces in buffer areas.

The State of Vermont has what it calls a *stormwater discharge permit fee* for new development. This fee system is two tiered and depends on the classification of the receiving water for the stormwater. The fee is charged for new development only; but once installed, there is an annual charge in addition to the one-time initial charge. For stormwater discharges into "Class A" waters, Vermont charges $1,170 for the initial permit and $255 per year thereafter. For discharges into "Class B" waters, Vermont charges $300 initially and $55 per year thereafter.

Now we return to Maryland where politics begin to take center stage in the stormwater debate, and fury begins to ensue.

In 2007, bills were introduced in both Houses of the Maryland General Assembly to establish an impervious surface fee on a statewide basis. The Maryland proposal followed the Vermont and Anne Arundel models in that it imposed the fee only on new development. The original version of the bill went further, in that it imposed a $2 per square foot fee on impervious surfaces in general but only $0.25 per square foot fees in preferred growth areas called Priority Funding Areas (PFAs).

The bill was extensively amended in the House of Delegates and then adopted as House Bill 1220. The bill was then sent to the Senate where it was not acted upon before the General Assembly adjourned. It died there. (House Bill 1220 contained an extensive list of how projected revenues were to be divided among existing state agencies and even to specific programs within such agencies. It also created a 17 member group called the Chesapeake and Atlantic Coastal Bays Green Fund Oversight and Accountability Committee to manage only the funds generated by this bill. The issues of revenue fragmentation and management diffusion are dealt with elsewhere in this book.)

As adopted by the House of Delegates, the 2007 bill had three revenue features. First, it set a one-time impervious surface fee at $0.50 per square foot on a statewide basis to be imposed when grading permits are issued, thus eliminating the preference for preferred growth areas and the penalty for building in non-preferred growth areas. The bill also imposed a series of flat fees on the square footage of impervious surfaces when building permits are issued. The flat fees were different for residential and commercial properties.

For residential property, there was a schedule based on the quantity of impervious surface. These fees were as follows:

Impervious Surface Area (sq. ft.)	Fee
<1,000	$100
1,000–2,500	$250
2,500–5,000	$500
5,000–10,000	$1,000
>10,000	$1,500

For commercial property, the fee was set at a straight $1 per square foot.

The fiscal note to the bill cautioned that the Maryland Department of Planning did not have comprehensive data upon which reliable revenue estimates could be made but, bearing that in mind, set out tentative revenue estimates of $62 million per year for the residential fees and about $22 million per year for the commercial fees, for a total of $84 million. (The Chesapeake Bay Foundation, on the other hand, estimated fees from this bill at $130 million; so there is apparently great leeway in these estimates.) Accurate revenue estimates are also complicated by an extensive series of exemptions provided in the bill as well as an "offset" program whereby developers could reduce their payments by up to 75% if acceptable offsets (creating less runoff) are installed.

In addition to the one-time fees imposed by this bill, Section 5 contains a provision requiring the Departments of the Environment and Planning to report to the Governor and the General Assembly on how a statewide system of impervious fees could be imposed on *existing* property. Thus, at least the House of Delegates was seriously considering the imposition of a two-tiered tax, much like Vermont's, on both existing property and new development.

In addition to the other revenue estimates mentioned above, the Chesapeake Bay Trust also prepared revenue projections based on data that they obtained from the Department of Planning. These figures broke down property developed in the sample year into nine categories based on both use and size. The percentage of impervious surface associated with each respective category was also listed. From these data, and from a study done by the Department of Planning in December of 2001, titled "Maryland's Changing Land: Past, Present and Future," some illustrative financial estimates were made bearing in mind all the above statements about sparse data and the inherent uncertainty involved in extrapolations.

That said, the Department of Planning report noted that there were approximately 1,145,000 acres of developed land in Maryland in 1997. It also noted that new development was in the 14,000-acre-per-year range. So, over the preceding 10 years, approximately 140,000 acres of land had been developed.

Going back to the Trust's data, the following categories of property constituted the following respective percentages of developed property and had the respective percentages of impervious surface per unit of land:

Category	% of Total	Developed	% Impervious	Acreage	Sq. Ft.
>2 acres	50	649,480	11	68,845	2,998,882,972
1–2 acres	14	175,630	14	21,115	1,094,013,320
0.5–1 acre	7	94,770	21	20,091	875,174,414
0.25–0.5 acres	7	89,830	28	24,973	1,087,812,554
0.125–0.25 acres	5	66,690	33	21,742	947,035,346
Townhouses	1	1,780	41	5,636	245,505,031
Institutions	6	77,090	34	26,519	1,155,165,898
Light industry	6	84,240	53	44,984	1,955,510,010
Commercial	4	48,360	72	34,916	1,520,937,475
				Total	**11,884,037,022**

With well over 11 billion square feet of impervious surface in Maryland, a flat impervious surface fee rate of only $0.01 per square foot would generate over $118 million. This sum, when coupled with the formula for new development contained in House Bill 1220, would result in some $200 million of annual fees. Assuming that the vast majority of these fees were held in one account and used to securitize 30-year bonds for Bay restoration, over $4.5 billion could be raised for grants for such projects.

Two final notes on the impact of these fees and the rationale for adopting them:

First, as noted above, the City of Takoma Park has a "stormwater fee" of $48 per ERU, which is defined as 1,228 square feet. Translating this fee into the same terms used to analyze House Bill 1220 and the estimates of such a statewide fee for existing property, the Takoma Park rate for residential property is about $0.04 per square foot of impervious surface. The impervious surface fee suggested above for existing property would add $0.01 per square foot, or $12.28 per ERU to this bill. Thus, for the residents of Takoma Park, this new fee would constitute a 25% increase. On the other hand, having a total rate of $0.05, of which 80% goes to deal with local stormwater run-off and the state share of 20% goes to deal with the larger issue of the receiving waters (to use Vermont's terminology), that is, the overall health of the Bay, seems a reasonable distribution of such costs. Furthermore, when viewed against other Bay restoration revenues such as the Bay Restoration Fund fee of $60 per household per year, a $12.28 household charge seems fair and reasonable.

Second, the issue of the rationale for adopting an impervious surface fee should be considered.

The local governments mentioned above that have imposed an impervious surface fee have done so to fund their local stormwater management programs. They have also done it to deter future development of land with large proportions of impervious surface. Maryland's proposal does deal with deterring future development of impervious surfaces, but only obliquely (largely through the local government retention section of the bill) deals with stormwater management.

Stormwater management is a local issue and restoration of the Bay is a state, if not regional, issue. Thus, the proposals described herein provide some funding for local government stormwater efforts and also do not impose so heavy a burden (as in the case with Takoma Park) that local governments are forestalled from imposing local future impervious surface fees of their own to further fund their stormwater management efforts.

Finally, as noted in the reference to the Vermont fee, those are based largely on the impact of run-off on the receiving waters, recognizing the State's role as steward of those waters. So, too, the Maryland proposals reflect the legitimate state concern of the impact that run-off has on the receiving waters of the Chesapeake and Atlantic Coastal Bays.

With that said, in 2012 the Maryland General Assembly abandoned its attempts to create a statewide impervious surface fee and, instead, turned its guns on local governments, specifically the ten jurisdictions that had received Municipal Separate Storm Sewer Systems (MS4) permits. These are the nine largest counties in the state, plus the City of Baltimore.

In sum, what the Maryland General Assembly did was to require each of these ten jurisdictions to create a "stormwater management agency" to deal with their local stormwater issues. More importantly, the General Assembly *required them to fund this new agency by enacting a specific new tax or fee to pay for their stormwater remediation initiatives.*

Notwithstanding a firestorm of adverse publicity about these "rain taxes," the City of Baltimore and seven of the nine counties complied in dutiful fashion with the new state law. But what is most telling about this controversy is what the other two counties did.

One of the counties dutifully created a local stormwater management agency but refused to fund it with a new tax or fee. They claimed that they will fund it out of general revenues. It remains to be seen how well this will go down during the 2014 session of the General Assembly. Will the General Assembly accept the county's word (and affront); or will they sue?

The other county's actions were even more outrageous. Oh, they dutifully created a stormwater management agency, and they also passed a new specific tax to fund the agency. This county has just fewer than 50,000 parcels of land under its jurisdiction. So, the county council enacted a tax of 1¢ per parcel! That is less than $500 for the whole county! It will cost far more to administer this tax than it will take in—and the county council knows this full well. The county council makes no bones about the fact that its actions are a message to the State.

And so, the issue drags on . . .

State Gasoline Tax

Per-capita gasoline consumption in the United States is 446 gallons per year. There are, as of the 2010 Census, 37,253,956 people in California. This means that the people of California use 16,615,264,380 gallons of gasoline. This means that a 1¢ increase in the gasoline tax would bring in $166,152,644, or enough to support bond issues for environmental projects of over $2.5 billion.*

Statewide Critical Area Tax Increment District/ Statewide Critical Area Special Tax District

These two fundraising options were discussed extensively in Chapter 6. Suffice it to say that, unlike extracting payments from polluters, these two revenue sources get their money from those who benefit from a high-quality environment. In the Maryland example, it is those whose property values are buoyed, if you will, by their proximity to the Chesapeake Bay. Taxing those who benefit is a relatively rare concept. It should certainly be developed.

Vehicle Emissions Inspection Program (VEIP)

Many states already have some form of vehicle emission inspection program—Washington, Connecticut, Illinois, Pennsylvania, Maryland, Arizona, and Rhode Island, to name a few.

To help pay for their environmental improvement programs, each state could increase their inspection fee by, say, $10, without too much pain.

As such, the State of Washington, which has 3,087,818 automobiles, would take in $30,878,180, which is enough to support the annual debt service payments on about $475 million of bonds for environmental improvement projects.

Of the states mentioned, Illinois has the most automobiles, with 5,947,468. They would take in $59,474,680, which would support almost $915 million of environmental improvement bonds.

Rhode Island has the least number of cars, with 508,389. But even Rhode Island could take in $5,083,890, which is enough to support over $78 million of bonds issued for environmental projects.

Water Use Tax†

States should consider imposing a *water use tax*, much akin to Maryland's Bay Restoration Fund "Fee." The theory upon which this proposal is based is

* California is probably the wrong state to use as an example. It already has the highest gasoline taxes in the country, at 39.5¢ per gallon.

† Maryland calls its "flush tax" a fee. But that didn't fool the experts; it is a tax. In general, the difference between a tax and a fee is whether payment can be avoided. If a payment can be avoided—by, for example, taking your garbage to the dump instead of having it collected—then that is a fee. If it cannot be avoided – If you are in Maryland and not on a sewer or septic system, what do you do when you flush your toilet?—then it is a tax.

similar to the thinking behind the Bay Restoration Fund Fee; that is, people should pay for the right to remove water from the watershed (water use fee) much as they should pay for contaminating the rivers and aquifers of the watershed (Bay Restoration Fund Fee).

For those receiving municipal water but who are not on meters, such a fee could be based on the average daily consumption of water, which in the United States is 100 gallons a day per person, or 250 gallons per household. The same is true if you are on a well. This means that the average household uses (in a 30-day month) some 7,500 gallons of water a month.

Now, Maryland's flush tax, for example, is $5 per month. Charging the same $5 a month for water use would be the same as charging .07¢, or $0.0007 per gallon.

Now while $0.0007 per gallon does not sound like much, $5 a month does. At least that was the reaction when Maryland enacted the flush tax, and then increased it.

If the water system is on a meter, then the water use tax could be structured as an Incremental Block Tariff to provide an escalating fee schedule depending on the quantity of water used. Thus, metered system users would pay the same $0.0007 per gallon for the first 7,500 gallons of water they used each month. But for any amount over and above 7,500 gallons, they would pay more per gallon.

In summary, there are many possible sources of revenue for renewable energy and environmental improvement projects. The key points to remember are that these revenues should be used to pay the annual debt service on environmental projects, and they should change or affirm behavior.

8

Financial Principles of Environmental Finance

It may seem redundant that these last eight principles of environmental finance are called the *financial principles;* but as you will see, they are just that: financial principles that should be observed when formulating environmental policy or organizing environmental finance programs.

Some of these financial principles are just common sense, but five of them are important enough and complex enough that they have their very own chapters in this book. The other three are also important, just simpler and shorter to explain.

Principle 16: Leverage the money. The most important word in environmental finance is *leverage*.

Many people think that *leverage* is a dirty word. They associate it with fiscal charlatans, or with the smoke-and-mirrors crowd on Wall Street. Nothing could be further from the truth. Oh sure, swindlers and crooks use leverage, but so do we all. Do you have a home mortgage? If you do, you have leveraged your down payment. Do you have an auto loan? If so, you have leveraged this down payment as well.

As you will see in Chapter 16, leverage, especially as it can be used in financial guaranties, is one of the most powerful concepts in environmental finance.

Principle 17: Stop general subsidies, which waste billions of dollars. Replace them with targeted subsidies for those who truly need them.

As you will see, we waste hundreds of millions of dollars each year on needless subsidies—subsidies for the wealthy.

Subsidies should be reserved for those who truly need them. In Chapter 1, you read about Jack and Diane who really needed a subsidy. Compare this to the benighted state policy that created a subsidy alright, but one that the truly needy could not avail themselves of—a subsidy that was largely wasted and policy that could just as easily have been implemented by regulation.

Chapter 15 deals with subsidies; it is titled "The Curse of Subsidies." You will see what I mean.

Principle 18: Make financial decisions based on strict cost/benefit analyses. Get politics and chance out of the decision-making process.

I hope you recall that the first, and most important, principle of environmental finance is to provide the greatest environmental benefit to the largest number of people at the lowest possible cost. How can we be sure that we are doing just that when we create an environmental finance program? The answer: By performing a cost/benefit analysis on each project in the program.

Chapter 18 sets out the rationale for, and methodology of, cost/benefit analyses for environmental projects. As you will see, there are some unique twists on how these analyses apply to environmental programs.

Principle 19: Never give grants unless absolutely necessary.

There are only four legitimate reasons for giving a grant: (1) because the recipient cannot pay for a necessity without it; (2) to induce someone to undertake an environmental improvement project that they are not legally required to do; (3) to develop, introduce, or commercialize a new environmental or renewable energy technology; and (4) for environmental education.

Chapter 13 deals with the concepts of grants and affordability.

In 1972, with the passage of the Clean Water Act (CWA), Congress created a "construction grant" program, which, over the next 15 years funneled over $70 billion in grants to major urban wastewater treatment facilities. In 1987, President Ronald Reagan persuaded the US Congress to abandon the construction grant program in favor of revolving loans. And so was born the Clean Water State Revolving Fund (CWSRF). To date, the CWSRF has provided another $100+ billion in financial assistance, 96% of which has gone to municipal wastewater systems.

What is the lesson here? We did not need the grants! Furthermore, all $70+ billion of those grant funds are gone for good.

On the other hand, as you read this page, the $100+ billion in the CWSRF is not only intact, but it is actually growing. Every state in the country, except Vermont, charges interest on the loans its CWSRF makes. These interest payments "revolve" back into the fund to be used for another project on another day.

Principle 20: Never make loans; guaranty them.

Now, having just extolled the virtues of the Clean Water Act's state revolving loan fund, let me say that, if given the option, you should never make loans; rather, you should guaranty them.

This may sound a little daft to you, but please reserve judgment until you have read the chapter (Chapter 16) on leverage that is subtitled "The Power of Guaranties." As you will see, you can create far more environmental benefits for far larger groups of people at a very low cost through the use of guaranties. You can actually create far more benefits—for the same money—than you can with loans.

Principle 21: Finance assets over the full term of their service lives.

This can sometimes be a very controversial principle. Many people believe that debt is evil. They believe that they should go through life without being financially beholden to anyone

Most people, however, cannot afford to pay cash for their home. So, the concept of a home mortgage becomes a necessary evil. So, too, with auto loans. Most people cannot afford to pay cash for their cars, so auto loans become a necessary evil as well.

These same people who buckle under and take out home mortgages usually do so for relatively short terms. They will choose a 15-year mortgage over a 30-year mortgage. They will do so even if they realize that in doing so, they have to settle for a much smaller house. If you can afford a $1,500-a-month mortgage payment, you can get a much bigger house with a 30-year mortgage than you can with a 15-year mortgage. For your $1,500 monthly payment, you can get only a $190,000 mortgage with a 15-year term. But with a 30-year term, the same $1,500 will get you about a $280,000 mortgage.

Now, what people do with their own money is definitely their own business. This book is not a clandestine attempt to beguile people away from their personal principles. So, as far as personal finances are concerned, eschewing debt is fine. But when it comes to public finance—to financing public facilities—it is rubbish.

The axiom to observe in public finance is that public assets should be paid for (i.e., financed) over the course of their useful lives. In public finance, the "useful" or "useable" life of an asset is generally called its "service life."

If you are on the school board and you need to finance a fleet of school buses, you are best advised to ask all of your transportation people—even vendors—how long these buses will last: How many years will they be able to ferry children back and forth between home and school. Let us say that the unequivocal answer you get is that the buses will last 10 years. In this case, your wisest course of action is to issue a 10-year bond to pay for the buses. You should not issue a 5-year bond. And you should definitely not issue a 20-year bond.

There is a principle at work here. It is a principle of environmental justice. It is called *intergenerational fairness,* by which is meant fairness to groups of people over time. Let me illustrate.

Let us say that you issue a five-year bond to pay for the school buses. This means that the taxpayers will pay double in the next five years.* Now, what about people who move into the district with children in year six? They will get the use of the buses for free for five years.

* We will use 0% (i.e., principal only) loans for this example.

Now, let us say that the board issues a 20-year bond. In this case, the taxpayers will pay half the cost of the bus over the next 10 years. Taxpayers who move into the district in year 11 will pay for the other half of the buses but they will get absolutely no benefit from their payments. Those buses are gone after 10 years, but their payments hang around for another 10 years. Not fair. Not good finance policy.

Here are four examples of finance policy involving the service lives of public assets:

1. When it comes to environmental infrastructure, most facilities have long service lives, in the 30- to 50-year range, and sometimes (as in the case of water and sewer mains) even longer. The US Department of Agriculture has been making 40-year loans to rural water and wastewater systems throughout the country since 1941. It is one of the most successful environmental programs on the planet.
2. Now recall that in 1987, Congress switched from a wastewater *grant* program to a wastewater *loan* program. When they did so, they limited the term of loans to wastewater systems to 20 years. They did so despite knowing about the success of the Department of Agriculture's program. Don't you think they would have, or should have, learned? Nope. The CWSRF has provided over $96 billion of 20-year loans to wastewater systems for facilities with service lives in the 30- to 50-year range!

   You know that the CWSRF is replete with subsidies, all aimed at lowering the cost of service. Well, if they would just extend the term from 20 years to at least 30 years, they could save enough money to get rid of some of those pernicious subsidies.
3. This example is even more egregious. It is another Maryland example. Do you remember in our discussions of the flush tax that we mentioned that the receipts from the tax are used to pay off bonds issued to finance Enhanced Nutrient Removal (ENR) projects? Well, Section 34 of Article III of the Maryland Constitution limits the pledging of tax revenues to 15 years. This is the same thing as limiting home mortgages to 15 years. You remember that for a $1,500 monthly payment and a 15-year term, you could get a $190,000 mortgage, whereas with a 30-year term, you could get a $280,000 mortgage for the same money. Well, right now, the $120 million that Maryland collects from the flush tax can support only about $1.4 billion in bonds. But if they amended the state Constitution to extend the pledge term to 30 years,* they could pay for about $2.1 billion of environmental projects. So, the people of Maryland—long-

* 30 years is the pledge term in Virginia, which is hardly a spendthrift state!

suffering with their flush tax and their new stormwater taxes—are leaving about $700,000,000 of environmental funding on the table, all because their stubborn political leaders refuse to extend the tax pledge term.

4. This final example is the most egregious of all. In the late 1970s, when the City of New York was in bankruptcy and the State was trying to extricate them, it was rumored in Albany that the New York City School Board had been financing pencils and paper—with service lives of only days or weeks—with 20-year bonds! Is it any wonder why the State imposed a Financial Control Board on the City?!

So, the rule is to finance public facilities over the course of their service lives. Not a day shorter—and definitely not a day longer!

Principle 22: Insist on full-cost pricing for environmental services. Full-cost pricing drives conservation and technological innovation. New technologies reduce cost.

As you will see in Chapter 15, titled "The Curse of Subsidies," environmental utility costs are heavily subsidized. This does no one any good. First, as you will see, if there are poor people who cannot afford their water or sewer bill, there are several strategies to effectively and compassionately deal with that.

Now the rest of the ratepayers who *can* afford the full cost of their service should definitely pay for it. No one likes to pay more. No board member or politician likes to raise rates. But rates can be raised gradually. (And when raising rates over, say, five years, the authorizing resolution should be passed today for all five forthcoming increases. This will save the board and/or politicians the anguish of having to go back to the people every year for more. Remember the flush tax example!)

Raising rates, whether to full-cost pricing levels, or not, will promote conservation. The higher the rate increase, the more people will conserve. Think of raising our gasoline rates to the $7 a gallon rates they charge in Europe. People would definitely find ways to drive less. Ditto when you raise water/wastewater rates. People would find ways to use less water.

Finally, higher rates will also drive innovation, which will have the eventual effect of lowering costs. Take the example of installing a technology that costs $10 million a system with full-cost pricing, that is, no subsidies. Engineers and scientists will know that if they can create a technology that does the same or better job at the same, or lower, cost, then they can get into the game without being trumped by some hidden subsidy.

Principle 23: Use good science and good economics to get accurate cost estimates. Low estimates cause anger and frustration; high ones cause denial.

In Chapter 7 you saw what was going on in Maryland with their stormwater problem. You may recall that one county simply refused to comply with a state law: They refused to create a new special tax to finance their stormwater projects, as they were legally required to do. And another county complied with only the letter of the law by levying a tax of one penny per parcel, a measure that would raise less than $500 for the entire county—an amount far below the cost of collecting it. This is what happens when local officials are faced with costs so high they have no real way of dealing with them: denial. And with this denial goes a bitter message to Annapolis that raises the heat on the whole stormwater issue.

On the low estimate side, Maryland has also had its problems. We have already mentioned Maryland's flush tax several times. Well, when they first decided to require their sixty-six largest sewage treatment plants to upgrade to ENR, they estimated the cost. Based on that cost estimate, they figured they would issue bonds, and then the flush tax would be used to pay the annual debt service payments on those bonds. So they passed the flush tax at a rate of $2.50 per month per household, which they figured would bring in about $60 million a year and would be enough to pay the annual debt service on the bonds they would issue to pay for all sixty-six ENR projects.

Unfortunately, as they started working on ENR implementation at the first few plants, they noticed that the costs were substantially more than they had originally estimated. Oooops! When they finished revising their estimates, they realized that they were not just off by a bit—they were off by several hundreds of millions of dollars! Oooops!

This meant that the General Assembly had to raise substantially more money to pay for ENR implementation. To do this, the General Assembly doubled the flush tax from $2.50 per month per household to $5 per month per household, or $60 per year per household. To say that this caused anger and frustration among the citizens of Maryland is an understatement. Their politicians told them they needed $2.50 a month to take care of the ENR issue, and then the politicians came back a couple of years later for another $2.50.

This is very bad policy—even if it was accidental policy. When asking for money from the people, it would have been better to wait until reliable estimates became available—that is what pilot projects are for. Two or three pilot projects could easily have been built so that the engineers could get a firm grip on the cost numbers. Then and only then should they have been presented to the people along with a revenue request.

* * * * * * *

So, now we have looked at all 23 principles of environmental finance. Over the next set of chapters, we will explain them in more detail and see how they fit into the greater scheme of financing environmental projects. Most importantly, we will see how they contribute to our overarching goal of creating the greatest environmental benefit for the largest number of people at the lowest possible cost.

9

Financial Mechanics

Most of the environmental finance programs on this planet involve grants, loans, or some type of tax credits. The majority, however, involve loans. So, to fully understand environmental finance so that successful policies can be made to accomplish environmental goals, we need to be able to deal with the mechanics of loans. There are several subjects that we must investigate in order to do so.

Before beginning, however, a word about bonds. A *bond* is nothing more than a loan that is on a standard form document where the borrower has a published credit rating. Think of a $1,000 US Savings Bond. All the language on all $1,000 bonds is the same. It is, of course, a standard form. When you buy a $10,000 bond from the State of Texas, which is part of, say, a $100,000,000 bond issue, you would find that each such bond has exactly the same wording.

In addition to standardized documentation, bonds also have published credit ratings, which are properly both the rating on that individual bond issue (that is, "bond issue," not individual bond) and to a great degree the rating of the entity issuing the bond. You can look up the credit rating on any bond you want to buy, or just ask your broker. So, a bond is just a loan in standard form with a known credit rating.

First we will examine the two most widely used types of loans: (1) level payment loans and (2) level principal payment loans.

Once we know the characteristics of these two types of loans, we then need to be able to compare them. There are two ways to compare loans. We will call them the *annual payment method* and the *total payment method.* These subjects will be discussed in Chapter 10. That chapter will also cover *discounting.* Essentially, in order to be able to compare loans by the total payment method, we must be able to discount their annual payments. The opposite of discounting is compounding. Compounding is useful to determine how much something will cost in the future.

The second concept we will look at in this chapter is how to determine *Cash Available for Debt Service* (CADS) and then how to use this number to determine several aspects of project financing, including affordability.

The third concept is *coverage.* If your CADS is $100, you cannot have an annual debt service payment (ADSP) of $100. No margin of error. The closer you cut to the bone, the worse will be the credit rating on your bond issue and the higher will be the interest rate. So, cutting coverage close is just shooting yourself in the foot.

Types of Debt

Types of debt are classified by their principal repayments. To understand the various types of debt, it is useful to construct an *Annual Debt Service Payment Schedule* for each particular type of loan.

An ADSP is the money paid back for a loan in a specific year. The list, or summary, of all the ADSPs of a particular loan is called its Annual Debt Service Payment Schedule.

There are four types of loans, or, four methods of repaying the principal of a loan:

1. Level payment method
2. Level principal method
3. Balloon payment method
4. Irregular payment method

Irregular Payment Method

The *irregular payment method* is only used where there is a spike or dip in the debt service payment stream. Let us say that you are running a water main down a street and connecting 300 houses. And you are going to charge each house $1,000 when they are all connected. Let us say that the project will take 2 years and cost $1 million. In such case, you might have interest-only payments in the first 2 years. Then in the third year, you have a $300,000 plus interest payment. Then in the next 17 years, you have level payments based on a $700,000 bond at the prevailing rate of interest with a 17-year term.

That is all that really needs saying about the irregular payment method. It is irregular because there is something irregular about the repayment stream. The irregular payment method is used only when circumstances require it.

Balloon Payment Loans

There are loans where the terms and the amortization periods are different. The key to the difference between the two is the phrase "scheduled to be." There are many loans where, for example, the principal is "scheduled to be" repaid over a 30-year period, but where the final principal payment is made at the end of year 10. Loans where the term and the amortization period are different are called *balloon loans*. As you might imagine, the final payment of such loans is called the *balloon payment*.

Some people have *balloon mortgages*. You might hear someone say, "I have a 30-year mortgage with a five-year balloon." This means that the loan has a 30-year amortization period and a five-year term. What this means to the mortgagor is that the first five annual payments are identical to the first five payments on a 30-year mortgage. (Remember: Mortgage loans are level payment loans.) At the end of year five, however, in addition to making annual payment #5, the hapless mortgagor must also pay the entire outstanding principal balance, which is then due and payable.

You might wonder what this poor soul's banker expects him to do at the end of the fifth year. How could anyone be expected to pay off 25 years of principal payments at one time? The answer, of course, is that no one expects the mortgage to be paid off at that time. It will simply have to be renegotiated. People often refer to this event as a *rollover*. Technically, it is not; nothing is actually rolled over. The old mortgage is, in fact, paid off and a new mortgage is created. The borrower will have to pay a different rate of interest. He may have a different term; and the amount of the new loan will be the outstanding principal balance of the old mortgage at the end of the fifth year. The *balloon* becomes the initial principal amount of the new mortgage.

This all may seem a little crazy; but actually it is not. Here is an example of how a balloon mortgage would work to a homeowner's advantage.

Let us say that a homeowner takes out a $300,000 mortgage at 5% interest with a 30-year term and a 10-year balloon. In such case, his annual payment* would be $19,515. Now, you know that in the first year of the loan, there is $300,000 outstanding and the interest rate is 5%. This means that the homeowner owes $15,000 of interest. However, his annual payment is $19,515. This means that the difference ($19,515 – $15,000 = $4,515) is principal. The homeowner actually pays off $4,515 of the $300,000 mortgage at the end of the first year. The table below shows how this looks over the 10-year balloon period.

Year	Principal Payment	Mortgage Loan Balance
1	$4,515	$295,485
2	$4,741	$290,743
3	$4,978	$285,765
4	$5,227	$280,538
5	$5,489	$275,049
6	$5,763	$269,286
7	$6,051	$263,235
8	$6,354	$256,882
9	$6,671	$250,210
10	$7,005	$243,205

As you can see, by the end of year 10, when the balloon payment is due, there is only a $243,205 balance. At this point, the homeowner does not pay off the $243,205 in cash; rather, he "rolls it over" into a new mortgage. Let us say that he gets another 30-year mortgage and that interest rates are still 5%. In this

* Although all mortgages are divided into monthly payments, for this example and for the sake of simplicity, I am going to use annual payments. The effect is the same. Using a monthly payment schedule would actually result in a lower balloon.

case, his annual payment on a $243,205 mortgage at 5% for 30 years is now $15,820. In other words, by taking advantage of the balloon payment mortgage, the homeowner wound up with an annual payment that was $3,694 *less* than his old mortgage payment.

Furthermore, if interest rates slid up to as much as 6.96%, he would still be paying the same $19,515 as on the old mortgage.

Finally, there is one more factor to consider: *buying power.* Let us say that the homeowner was a prudent person who did not want his mortgage payment to exceed more than 20% of his income. So, let us say that the homeowner's income was $100,000, making his $19,515 annual mortgage payment just a bit under his self-imposed 20% limit.

But earnings are not static. It is highly unlikely that our homeowner earns the same today that he did 10 years ago. Indeed, let us say that his income increased by a modest 2% a year. Thus, his $100,000 income 10 years ago should now be $121,899. If he still had his self-imposed annual mortgage payment limit of 20%, then he should be willing to pay as much as $24,380 for his new mortgage. This means that by using the same proportion of his income ($24,380, or 20%), he could now afford an interest rate of 9.7%. Quite a jump from 5%!

So, as you can see, balloon mortgages are not crazy at all. But you do have to be a little bit adventurous!

So, there are essentially two types of loans. The first, and by far the more prevalent, are loans where the term and the amortization period are the same. The second, which you just saw in action above, is the balloon loan where the amortization period exceeds the term. As it turns out, balloon mortgages are much more prevalent in the commercial mortgage market than they are in the residential market. This is for the very good reason that most commercial real estate is leased. Most leases contain rent escalator clauses. Thus, commercial property owners are better equipped financially to deal with the vagaries of what the new interest rates will be when they roll over their balloon.

Please note that there is no such thing as a loan where the term exceeds the amortization period. By definition, once the principal is repaid, there is nothing left upon which to charge interest, and the loan is over.

Level Payment Method

In the United States, most municipal utility debt is level payment debt. This is very suitable for utilities that have fixed rates (tariffs) that they charge their customers, as the payments are the same each year. In Europe and Asia, the level principal method is used more often. Because the ADSPs decline each year with the level principal payment method and because tariffs are fixed, utilities wind up with considerable extra revenue over time.

The level payment method means that *the total amount of principal plus interest paid is the same every year.* As the amount of interest declines with every payment that is made, the amount of the principal paid increases by the same amount.

Calculating the *ADSPs* for level payment loans involves the use of a complicated formula, as follows:

$$\text{ADSP} = \text{P} \times (\text{i}/(1 - (1/(1 + \text{i})^{\text{n}})))$$

Handheld business calculators can deal with this formula very easily using four keystrokes. But for those without such devices, the ADSP on a 40-year loan can take a long time to calculate. Just to illustrate how the formula works, here is the calculation of a 2-year loan of $100 at 5% interest. There are seven steps involved. Here is the legend:

ADSP: Annual debt service payment
P: Original principal amount of the loan
i: Interest rate expressed as a decimal
n: Term of the loan

Here are the seven calculations:

1. $\text{ADSP} = \text{P} \times (\text{i}/(1 - (1/(1 + \text{i})^{\text{n}})))$
2. $\text{ADSP} = \$100 \times (0.05/(1 - (1/(1 + 0.05)^2)))$
3. $\text{ADSP} = \$100 \times (0.05/(1 - (1/(1.05)^2)))$
4. $\text{ADSP} = \$100 \times (0.05/(1 - (1/1.1025))$
5. $\text{ADSP} = \$100 \times (0.05/(1 - 0.90702))$
6. $\text{ADSP} = \$100 \times (0.05/0.09298)$
7. $\text{ADSP} = \$100 \times 0.53775$

$$\text{ADSP} = \$53.76$$

You can easily prove that this is correct. You know that 5% interest on $100 is $5. So, of the first year's payment of $53.76, you know that $5 of this amount must be interest and the rest, $48.76, must be principal. If you pay off $48.76 of principal in the first year, there is only $51.24 of principal outstanding on the loan at the beginning of the second year. The second year's payment must be $51.24 of principal plus 5% interest. Five percent of $51.24 is $2.52. So, the principal of $51.24 and interest of $2.52 equals the ADSP of $53.76. So, now you know that the formula is correct.

In going forward, for the purposes of illustrating the various principal payment methods, let us assume a 5-year loan of $100,000 with an interest rate of 10%. To find the ADSP, we simply plug the variables into the formula:

$$\text{ADSP} = \$100{,}000 \times (0.10/(1 - 1/(1.10)^5))) = \$26{,}380$$

Once we have the ADSP, we can now create an Annual Debt Service Payment Schedule to determine exactly how much interest and principal are paid each year. An Annual Debt Service Payment Schedule is also called an *Amortization Schedule.*

Here are the rules for creating an amortization schedule for the level payment loan in our example:

Creating an Annual Debt Service Payment Schedule for the Level Payment Method

1. Determine the ADSP using the equation above.
2. Calculate the first year's interest payment: (Interest = i × Principal).
3. Obtain first year's annual principal payment by subtracting the interest from the ADSP of the first year: (Principal = ADSP – Interest).
4. Subtract the year 1 principal payment from the original principal amount.
5. Calculate the second year's interest payment (Interest = i × Outstanding Principal Balance).
6. Repeat the above process for each year of the loan term.

The table below shows what this process looks like in tabular form.

Year	Prior Balance	Interest	–	Total Annual Payment (ADSP)	=	Principal Payment
1	$100,000	$10,000	–	$26,380	=	$16,380
2	$83,620	$8,362	–	$26,380	=	$18,018
3	$65,603	$6,560	–	$26,380	=	$19,819
4	$45,783	$4,578	–	$26,380	=	$21,801
5	$23,982	$2,398	–	$26,380	=	$23,982
	Total	**$31,899**	–	**$131,899**	=	**$100,000**

Summary of the Level Payment Method

- The annual payments are always the same.
- The amount of principal paid each year increases by the exact amount in which the interest payment decreases.
- The total of all the annual principal payments equals the original principal amount of the loan.

Level Principal Payment Method

The level principal payment method means that the total amount of principal is the same from year to year, and the interest payment is simply calculated using the remaining balance. As the amount of interest declines with every payment that is made, the total annual debt service payment also decreases.

When calculating ADSPs for loans using Level Principal Payment Method:

P: Original principal amount of the loan

n: Term of the loan

Notice that the annual principal payment is the original principal amount divided by the term (P/n).

In the case of the example of a five-year loan of $100,000 at 10% interest, the annual principal payment would look like the following:

Annual principal payment = $100,000/5 = $20,000

The annual debt service payment would be different in each year. But in the first year, it would be the interest payment, which is 10% of the original principal balance of $100,000, or $10,000, plus the annual principal payment of $20,000. So the first year's ADSP would be $30,000.

The table below shows what the full amortization schedule would look like in tabular form.

Year	Prior Balance	Interest	+	Principal Payment	=	Total Annual Payment (ADSP)
1	$100,000	$10,000	+	$20,000	=	$30,000
2	$80,000	$8,000	+	$20,000	=	$28,000
3	$60,000	$6,000	+	$20,000	=	$26,000
4	$40,000	$4,000	+	$20,000	=	$24,000
5	$20,000	$2,000	+	$20,000	=	$22,000
	Total	**$30,000**	+	**$100,000**	=	**$130,000**

Here are the rules for creating an amortization schedule for level principal payment debt.

Creating an Annual Debt Service Payment Schedule for the Level Principal Payment Method

1. Calculate the level annual principal payment by dividing the original principal amount by the number of years in the term of the loan (annual principal payment = P/n).

2. Calculate the first year's interest payment by multiplying the original principal amount of the loan by the rate of interest: (Interest = P × i).
3. Obtain the first year's annual payment by adding the level principal payment to the first year's annual interest payment: (1st year ADSP = Interest + Principal payment).
4. Obtain the outstanding principal balance at the end of the first year by subtracting the first year's level principal payment from the original principal amount of the loan: (Outstanding Balance = P – ADSP).
5. Calculate the next year's annual interest payment by multiplying the outstanding principal balance from the preceding year by the rate of interest: (2nd year interest = Outstanding balance × i).
6. Calculate the next year's annual payment by adding that year's annual interest payment to the level principal payment: (2nd year ADSP = 2nd year interest + Principal payment).
7. Continue the process for each year of the term.

Summary of the Level Principal Payment Method

- The annual principal payments are the same each year.
- Both the annual interest payment and the annual payment decrease each year.
- The total of all the annual principal payments equals the original principal amount of the loan.
- Note that if there were no interest (an interest rate = 0%), then the level principal method would be the same as the level payment method, that is, the payments would be the same each year.

Cash Available for Debt Service (CADS)

It may seem strange, but Cash Available for Debt Service (CADS) plays a more important role in environmental finance than it does in corporate finance. In corporate finance, CADS is the limit of what you can pay for a project. So, if a company's CADS is fixed (which it almost always is) and the company needs to undertake a large, expensive project, then the finance director has to scour the Earth for a loan with the longest possible term and the lowest possible rate—a loan whose ADSP is below his CADS. But at the end of the day, if the company cannot find a loan that they can cover with their CADS, then they just cannot undertake the project they want.

Such is not the case when dealing with environmental finance, especially when dealing with drinking water, wastewater, toxic air pollution, or even solid waste projects, which must simply get done—for public health reasons—regardless of the cost. In these cases, CADS plays a more robust role in the financing.

CADS can be defined as "recurring revenues less cash expenses." The first step in calculating the CADS is identifying recurring revenues, or Regular Cash Income. Regular Cash Income encompasses all monies that are collected from the *daily* operations and management of the utility. The term *regular* does not mean ordinary, but instead means predictable and continuous. Tariffs and interest that have proven to be constant and reliable sources of income can be projected Regular Cash Income in the future. Non-Recurring Cash Income cannot be used to predict future income and therefore cannot be counted on as cash available for debt service. Non-Recurring Cash Income includes such things as the income from the sale of assets such as land or equipment, government grants, and user fees. Only in short-term cases of very constant growth can connection fees and similar fees for service be considered Regular Income.

This is important. Utilities cannot count one-off income events, such as from the sale of a piece of land, as recurring income. Basically, recurring income is the fees collected from the ratepayers plus any interest on any savings accounts.

Environmental utilities should not have much CADS hanging around. Their rates should be set so that they can pay their bills and put away a little bit in a savings account for emergencies. So, when undertaking a major project—one that will involve a rate increase—the utility management counts the CADS and uses that number as an ADSP to estimate how large a project they could undertake without raising rates.

Determining the size of a project requires a complicated use of our favorite formula: $ADSP = PV \times (i/(1 - (1/(1 + i)^n)))$. In this case, the ADSP is our CADS. We need to solve for the PV, which is the principal amount of the loan necessary to finance our project.

What is missing here, of course, are both the "i," the interest rate, and the "n," which is the term. So the finance director must search through all applicable federal and state environmental finance programs to see if his project is eligible for any such programs. He also needs to talk to his financial advisor about the availability of the municipal bond market. If he can find a source of funding, he then needs to find out at what interest rate the programs are lending and, most importantly, for what term.

Now let us say that the utility has CADS of $100,000. Let us say that the finance director has identified a program for which his project is eligible that will lend at an interest rate of 4% for a term of 20 years. Now the finance director can plug these variables into his financial calculator and determine that his $100,000 of CADS, used as an ADSP, can support a 20-year loan at 4% of $1,359,033, call it $1.36 million.

Now let us say that the finance director sits down with his boss and his board to talk about rate increases. In this case, the board members overwhelmingly feel that their ratepayers can only afford a rate increase that will yield another $100,000 of additional income to the utility.

So now the finance director goes back to his calculator and plugs in $200,000 as the ADSP. Using the same rate and term, he calculates that with a $200,000 ADSP, his utility can afford a $2.72 million project.

The bad news comes when the utility's consulting engineers tell the management and the board that the project simply cannot be done for a penny less than $4 million. This means that the utility is short $1,280,000. They need a grant for $1,280,000; otherwise the project does not get done. If this system is a rural water system, the US Department of Agriculture (USDA) has a rural water and wastewater program that provides both grants and loans, so our utility might have to wait a year but the project can eventually get done.

If our utility is in a town of more than 10,000 people, then it is ineligible for the USDA program. There are no other federal grant programs for environmental utilities. So, unless the state has a grant program, or the governor and legislature of the state can be persuaded to enact a bill to provide a grant, the utility is just out of luck. In this case, they would either have to abort the project or go back and "bite the bullet" and raise the rates to cover another $1,280,000 of debt.

If that news is not bad enough, there is more.

Coverage

In the above examples, we have been quite glib in saying that, for example, with $200,000 of CADS, you can get a $2.72-million loan for 20 years at 4%. No banker or investment banker would let you enter into this transaction. Why? No margin of error. In the finance business, your margin of error is called your *debt service coverage*, or just your *coverage* or your *coverage ratio*. No one will give you a loan or buy your bond if all you have is a 1:1 coverage ratio. This means that you must have more CADS than you need.

The question is: How much more? The answer to this question will lead you and your finance director into a very useful exercise.

If you are dealing with banks, you can ask them and they will tell you. If you have the luxury of having more than one bank willing to lend to you, then you can ask both. Then you can compare both loans, factor in the coverage necessary to satisfy them, and determine what it will mean in terms or increases for your ratepayers.

A better and more instructive way of ascertaining necessary coverage is to consult one or more of the three international credit rating agencies. If you are going to seriously consider issuing a municipal bond, you will have to do so anyway.

The three international credit rating agencies are Standard & Poor's, which is owned by McGraw Hill Financial; Moody's Investors Service; and Fitch Ratings. Using Standard & Poor's ratings scale, investment-grade ratings go from AAA to BBB–. Anything BB+ or below is called a *speculative* investment or a *junk* bond.

Assuming your utility is financially and managerially healthy and is situated in a reasonably prosperous and upbeat area, you should clearly be looking at an investment-grade rating on your bonds. Now, here is where the fun starts. Standard & Poor's will eventually look at your coverage ratio. If it is in the 1.1:1 range, there will be a lot of tsk-tsking and talk about maybe a BBB- rating*. If you start talking about a 1.25:1 coverage ratio, the conversation may brighten to a discussion of an A rating. Talking about a 1.5:1 coverage ratio might get you into the AA category, whereas a 2.0+:1 coverage ratio might get you definite consideration for a AAA rating.

So what is this all about? What is the value of a AAA rating? The answer is that the higher the rating, the lower the interest rate you will pay on your bonds. Is there any magic formula? Should your utility always strive for a AAA rating regardless of what it means to your ratepayers? The answer is no. Credit ratings, coverage ratios, interest rates, and water or sewer rates are all trade-offs.

For example, let us say that the rating agency you are talking to you actually tells you that they will rate your bonds AAA if you will provide a 2.0:1 debt service coverage ratio. Let us say that they also tell you that they will rate your bonds AA if you will maintain a 1.5:1 debt service coverage ratio. In this case, the appropriate action for your finance director is to consult the interest rate paradigm that is published daily by all the financial reporting services. They will tell you and your director exactly what the interest rate is for a AAA bond and a AA bond, at whatever term you choose. Then the finance director can make his decision and his recommendation. He may determine that the interest rate savings is so great that a AAA rating is actually needed despite the pain of a higher increase for the ratepayers. Or, he may conclude that the interest rate on a AA bond—although a bit higher than that on a AAA bond—is far better than having to raise the extra money involved in a 2.0:1 coverage ratio.

As we said before, all things being equal, credit ratings/interest rates are a trade-off with debt service coverage ratios. As a utility board member, you should be aware of this and use the analysis to make the right decision for your people.

* These relationships between coverage ratios and credit ratings are purely my own, which I only use for illustrative purposes. They are certainly not Standard & Poors' material.

10

Comparing Financing Alternatives

This chapter will deal with comparing loans or bonds. As you know, a bond is simply a loan on standardized documents with a known credit rating. First, we will demonstrate how to compare simple loans. Then we will look at how to compare loans when there are other fees or charges involved. These are like the "points" that the banks charge on home mortgages as well as attorneys' fees, appraisal fees, servicing fees, and the like. But first we will start by comparing the two most common forms of loans: level payment loans and level principal payment loans.

Comparing Loans

There are two ways of comparing loans or bonds: (1) the annual payment method and (2) the total payment method.

Annual Payment Method

Recall now that there are two major types of loans: level payment loans and level principal payment loans. Well, in terms of comparing them, the annual payment method only works with level payment loans. But it is very powerful.

The annual payment method consists of calculating the annual payments on the level payment loans you are working with, and then comparing them. Very simple.

The Annual Debt Service Payment (ADSP) on a level payment loan (Loan A) of $10,000,000 at 6% for 10 years is $1,358,680. The ADSP on a level payment loan (Loan B) of $10,000,000 at 4% for 5 years is $2,246,271.

Loan A = $1,358,680

Loan B = $2,224,271

Certainly there is no problem comparing these two loans by the annual payment method. The next question is: If you were a member of your water authority's board, which one would you choose? What if your system had needed a major rate increase in the past year or so, and now you are faced with this—an additional rate increase? Would you choose the $1.3-million rate increase or the $2.2-million increase? Easy question, right? An important part of your job on the board is to hold down the costs your customers have to pay. Does it concern you that Loan A is 5 years longer than Loan B? Probably not. Does it concern you that the interest rate on Loan B is 50% lower than the rate on Loan A? Again, probably not.

As you can see, the annual payment method of comparing loans is very useful, very powerful, and, fortunately, very simple.

Total Payment Method

The total payment method looks easy, but it is not. You would think that all you would have to do is add up the annual payments on two loans and then look at them. So, in the example above, we would just multiply the $1,358,680 level annual payment in Loan A by its term of 10 years, or $13,586,800. Then we would multiply the $2,224,271 level annual payment in Loan B by its term of 5 years, or $11,121,355. You would then compare them, right?

Loan A = $13,586,800

Loan B = $11,121,355

Comparing them this way is easy, isn't it? Yes. But it is wrong. This is not a comparison of the total payments of these two loans. We have just compared apples to oranges.

Unfortunately, in order to correctly compare any two loans, you must discount each of their annual payments. *Discounting,* and its directional counterpart, *compounding,* are the two elements of the *Time/Value Theory of Money.*

Time/Value Theory of Money

We must begin by pointing out that the Time/Value Theory of Money is not a theory at all. It is fact. Absolute fact. The value of money actually changes over time.

Now we are entering another of these financial areas that is fraught with superstition, much like the concept of leverage. Everyone basically realizes that things cost less years ago. They even begrudgingly admit that things will cost more in the future. But how it all works is a bit of a mystery. Measuring the increases seems somehow like black magic.

Let me say this. Measuring back in time is easy. If you know for certain how much an object cost a specific number of years ago and you know exactly how much it costs today, then you can calculate the discount rate going backward or the compounding rate going forward. These, of course, are both the same.

Measuring forward, however, is another matter. Going backward, you can *find* the discount rate. Going forward, you *do not know* the rate. Those who claim great accuracy in predicting future costs and prices must, indeed, be practicing the black arts.

Most costs rise at, or close to, the rate of inflation. Some costs (college tuition, and medical care, for example) seem to rise at rates greater than that of inflation. Importantly for us, so do construction costs. But, in general, as we said, most prices rise at, or about, the rate of inflation.

The rate of inflation is compiled by the US Bureau of Labor Statistics from observing actual prices of a large number of goods and services. From January 2012 to late in 2013, the rate of inflation in the United States was well below 3%. But, as the Bureau reports, the average rate of inflation in the United States from 1914 to 2013 was 3.35%. So, let us call it 3% for our purposes.* Hyper-exactitude is unnecessary for the illustrations we are doing here. No one's life depends on our work.

As far as the Time/Value Theory of Money is concerned, the word "theory" has nothing to do with *whether* the value of money changes over time; rather, it only has to do with *how* the value of money changes over time. That said, we are now going to present a series of three examples to illustrate just how the value of money does change over time.

The first, and perhaps most familiar, is the savings account. The second example deals with a US Savings Bond and how the surrender value is calculated. And the third example uses a 30-year fixed-rate mortgage to show the subtle changes, which are actually occurring, when neither the home nor the payment appear to change at all over the 30-year period.

Next we will present some actual illustrations of how the change in the value of money over time relates to the payments a water or wastewater system makes on behalf of its projects.

Now, for the first example: the savings account. Let us say that you have $10,000 that you want to put into a safe investment account to have for your retirement in 10 years. Let us say that you find a fixed-income mutual fund that will offer you a guarantied rate of 5%, compounded annually. How much money will you have in 10 years?

(Before we go any further, please note that the terms *present value* and *future value* are relative terms. They are relative to each other. *Present value* refers to the value that is earlier in time. *Future value* refers to the value that is later in time. So, when we calculate the price of a quart of milk 50 years ago, today's price is the future value, and the price 50 years ago is the present value.)

---

* Corporations often use their own discount number that they estimate as their profit goal. They sometimes call this their *opportunity cost* or *hurdle rate*.

Here is how we calculate the value of your retirement account in 10 years:

\$10,000	
× 1.05	(1)
\$10,500	
× 1.05	(2)
\$11,025	
× 1.05	(3)
\$11,576	
× 1.05	(4)
\$12,155	
× 1.05	(5)
\$12,763	
× 1.05	(6)
\$13,401	
× 1.05	(7)
\$14,071	
× 1.05	(8)
\$14,774	
× 1.05	(9)
\$15,513	
× 1.05	(10)
\$16,289	

There you have it. Your \$10,000 retirement account will have grown to \$16,289 in 10 years at a *compound* rate of 5% per year. There is a formula for this exercise:

$$FV = PV \times (1 + r)^n$$

This is the *compounding formula*. Here, FV is, of course, the future value and PV the present value; the r is the interest rate and n the term of years.

The opposite of the compounding formula is the *discounting formula*:

$$PV = FV/(1 + r)^n$$

This formula can also be expressed as follows:

$$PV = FV \times (1/(1 + r)^n)$$

In other words, getting from the original \$10,000 to the future value of \$16,289, you multiply the amount by 1 plus the interest rate, each year. Getting back from the future value of \$16,289 to the present value of \$10,000, you simply take the \$16,289 and divide it by "1 + r" each year.

* * * * * *

Now let us take an example of a US Savings Bond. This is what happens when someone gives a baby a $1,000 US Savings Bond to help pay for the baby's college education. US Savings Bonds mature in 20 years. So, we know we can redeem the bond for $1,000 in 20 years. What is it worth today?

Well, using the second discounting formula, we would first add the rate of interest (0.05) to the number one (1) to arrive at 1.05. Then we would divide the number one (1) by 1.05 to get 0.9524. Then we would multiply $1,000 by 0.9524 twenty times. The answer, which is the value of the bond today, is $376.89.

* * * * * *

Our third and final example involves a 30-year home mortgage. There are two lessons in this example.

Let us say that you took out your 30-year, fixed-rate mortgage in 1990 for $150,000 at a rate of 6% on a home you purchased for $190,000. As such, your first monthly payment in 1990 was $899.33, call it $900. Let us also say that you were earning $50,000 a year back then. This means that your yearly house payments (12 × $900 = $10,800) equaled about 20% of your gross salary of $50,000.

Now, your last payment in 2019 will also be $900; but by then you should be making about $121,000 a year.* Your salary went up, but your house payment remained the same. So, in 2019, your yearly house payments will only comprise about 9% of your gross salary. This is lesson number one. Here the value of your work (i.e., as it is reflected in your salary) changed. It grew over time. And as it grew, the cost of your home, which was fixed, became a relatively smaller and smaller part of your earnings. So, as your earnings grew, the burden of your house payments shrank. This is illustrative of how the value of money changes over time. And you can measure it against any sum that is fixed over time, in this case your mortgage payment.

The second lesson involves the value of your home. Let us say that in 2019 the kids will be grown and gone, and it will be time to sell the house and downsize. So, you will sell your home in 2019.

If the value of your home inflated at the same pace as the growth of your salary, then in 2019 it should be worth about $461,000.

Now you should be familiar with the concept that the value of money changes over time. So, you should not be surprised to learn that in 1990 dollars, your final mortgage payment is not worth $900 at all. In fact, it is worth only $149.44. This is just like the example of the US Savings Bond. There we knew what the future value was ($1,000), so we could calculate back to the present value ($376.89). Here, too, we know the future value of the final

* Compounding and discounting throughout the chapter at a rate of 3%.

payment. It is $900. So we can calculate back to determine its value 30 years before, which is $149.44.

* * * * * *

Now let us see how all of this applies to the financing of environmental infrastructure.

Let us say that that your water system, which you are on the board of, has a $1,000,000 project. Let us say that you have three friendly local bankers who would like to lend you the money:

- Banker A will lend to your system for 5 years at a rate of 8% with level principal payments.
- Banker B will lend to your system for 5 years at a rate of 8% but with level payments.
- Banker C will lend to your system for 10 years at a rate of 9% and also with level payments.

Let us look at these loans. Here are the annual payments on Loan A:

Year 1	$280,000
Year 2	$264,000
Year 3	$248,000
Year 4	$232,000
Year 5	$216,000

To obtain the annual payment on Loan B, we must use the formula we saw above because it is a level payment loan:

$$\text{ADSP} = \text{P} \times (i/(1 - (1/(1 + i)^n))),$$

which in our case is

$$\text{ADSP} = \$1{,}000{,}000 \times (0.08/(1 - (1/(1 + 0.08)^5))) = \$250{,}456.$$

To obtain the annual payment on Loan C, we use the same formula, only in our case it is

$$\text{ADSP} = \$1{,}000{,}000 \times (0.09/(1 - (1/(1 + 0.09)^{10})))$$

So, the annual payment on Loan C is $155,820.

Let us stop for a moment and compare* these three loans using the annual payment method of comparison:

* You can only compare loans of the same principal amount. But you can do so regardless of differing interest rates or differing terms. However, you cannot compare a $10,000 loan with a $12,000 loan.

Loan A: \$280,000*
Loan B: \$250,456
Loan C: \$155,820

Is there any doubt in your mind as to which loan you will choose? There is no impeachment procedure that I know of for water authority board members; but if you voted for any loan other than Loan C, I am certain that at least some of your ratepayers would try to impeach you—or at least call for your resignation.

When you vote for Loan C, you are making a political decision. Nothing wrong with that. It is certainly your duty to contain the costs that your ratepayers must pay. What would be wrong, however, would be if you made that decision in ignorance of the true cost of these loans. To learn this, we must compare them by the total cost method. And to do so, we need to discount each of their annual payments. To do this, we will use a discount rate of 3%, and we will use the discounting formula: $PV = FV/(1 + r)^n$. Here, of course, FV is the annual payment in any given year, r is the interest rate, 8% for Loans A and B and 9% for Loan C, and the n is *the year of each respective annual payment*. So, we have to do the formula five times, once for each annual debt service payment.

Here are the discounted values of the annual payments for Loan A:

Year	**ADSP**	**PV**
Year 1	\$280,000	\$271,845
Year 2	\$264,000	\$248,845
Year 3	\$248,000	\$226,925
Year 4	\$232,000	\$206,129
Year 5	\$216,000	\$186,323
Total[a]		**\$1,140,067**

[a] Adding up the ADSP column is nonsense. They are all apples and oranges.

Now we can discount the annual payments on Loan B, as follows:

Year	**ADSP**	**PV**
Year 1	\$250,456	\$243,161
Year 2	\$250,456	\$236,079
Year 3	\$250,456	\$229,203
Year 4	\$250,456	\$222,527
Year 5	\$250,456	\$216,046
Total		**\$1,147,016**

* Please note that this is just the first year's payment. As you know, with the level principal payment method, the annual payments decline each year.

Now we obtain the present values for each of the annual debt service payments in Loan C, as follows:

Year	ADSP	PV
Year 1	$155,820	$151,282
Year 2	$155,820	$146,875
Year 3	$155,820	$142,597
Year 4	$155,820	$138,444
Year 5	$155,820	$134,412
Year 6	$155,820	$130,497
Year 7	$155,820	$126,696
Year 8	$155,820	$123,006
Year 9	$155,820	$119,423
Year 10	$155,820	$115,945
Total		**$1,329,177**

So now comparing these three loans by the total payment method, we have the following:

Loan A: $1,140,067

Loan B: $1,147,016

Loan C: $1,329,177

As you can see, Loan A, which is the most expensive loan when comparing by the annual payment method, is the least expensive loan when comparing by the total payment method. Conversely, Loan C, which is—by far—the least expensive loan when comparing by the annual payment method, turns out to be the most expensive—by far—loan, when comparing by the total payment method.

So, as a board member of the water authority, you can certainly vote for Loan C, because it will cost your ratepayers the least, even though they may have to pay more over time. When casting your vote, however, please remember to be sure that the service lives of the assets you are financing are at least 10 years. Otherwise, you will be violating Environmental Finance Principle 21.

Let us now look at another example, this time when the inflation rate is in a state of flux, or where the project has a heavy labor component and the rate of inflation for labor is higher than for the norm.

Let us say that an environmental utility manager is facing a 4% to 6% rate of inflation because of construction cost increases and wants to measure his total project costs at two different rates, for comparison purposes. Let us say he has the following options for a $1,000,000 project:

Option A: 5% with a 5-year term (no points/fees)*

Option B: 6% with a 10-year term (no points/fees)

Both loans are level payment loans.

The level annual debt service payment for Option A can be obtained by using our formula with the appropriate variables:

$$ADSP = \$1{,}000{,}000 \times (0.05/(1 - (1/(1 + 0.05)^5)))$$

The level annual debt service for this loan is $230,975.

We can use the same formula to determine the level annual debt service payment for Option B:

$$ADSP = \$1{,}000{,}000 \times (0.06/(1 - (1/(1 + 0.06)10))) = \$135{,}868$$

So, if we were only interested in comparing these two loans by the annual payment method, then we would have

Option A: $230,975

Option B: $135,868

This looks like a no-brainer for our environmental utility executive; but there is more to it.

As we said above, however, our executive is confronted by a range of inflationary factors between 4% and 6%. So, to be sure, he needs to discount each of the annual payments at both 4% and 6%. Here is Option A discounted at 4%:

Year	**ADSP**	**PV**
Year 1	$230,975	$222,091
Year 2	$230,975	$213,549
Year 3	$230,975	$205,336
Year 4	$230,975	$197,438
Year 5	$230,975	$189,845
Total		**$1,028,259**

* Points and fees will be taken up below.

Now, here is Option A discounted at 6%:

Year	ADSP	PV
Year 1	$230,975	$217,901
Year 2	$230,975	$205,567
Year 3	$230,975	$193,931
Year 4	$230,975	$182,954
Year 5	$230,975	$172,598
Total		**$972,951**[a]

[a] Note that the PV of the annual payments of Option A is lower than the original principal amount of the loan. This is because the interest rate on the loan is below the rate of inflation that the utility executive is using to determine his true cost. This means that the bank must be using 3%, or maybe lower, as the rate of inflation that they see. Otherwise, a banker will soon be fired, or his bank will close its doors for good.

Now let us look at Option B. First let us discount loan B's annual payments at 4%:

Year	ADSP	PV
Year 1	$135,868	$130,642
Year 2	$135,868	$125,618
Year 3	$135,868	$120,786
Year 4	$135,868	$116,141
Year 5	$135,868	$111,674
Year 7	$135,868	$103,249
Year 8	$135,868	$99,277
Year 9	$135,868	$95,459
Year 10	$135,868	$91,788
Total		**$1,102,012**

Now let us look at Option B's level annual payments discounted at a rate of 6%:

Year	ADSP	PV
Year 1	$135,868	$128,177
Year 2	$135,868	$120,922
Year 3	$135,868	$114,077
Year 4	$135,868	$107,620
Year 5	$135,868	$101,528
Year 6	$135,868	$95,782
Year 7	$135,868	$90,360

Year	ADSP	PV
Year 8	$135,868	$85,245
Year 9	$135,868	$80,420
Year 10	$135,868	$75,868
Total		**$1,000,000**[a]

[a] For those of you checking the work, I was rounding. The numbers actually add up to $999,999. When the interest rate on a loan equals the particular discount rate you are using, the answer is always the original principal amount of the loan.

So, now, here is what both Option A and Option B look like under both inflation rate scenarios:

Inflation Rates	4%	6%
Option A	$1,028,259	$972,951
Option B	$1,102,012	$1,000,000

As you can see, in each instance, Option A offers the utility executive the lowest cost, regardless of either discount rate. It would be especially good for his system if his costs (and his revenues) were increasing at 6%. Then the loan would actually make the system a modest profit, odd as that may sound. (Later in this book, we will be discussing grants and affordability and we will run into grant situations that are triggered when the rate of inflation exceeds the interest rate on a given loan. But, again, this is very rare. This means that the bankers are losing money and the borrowers are making money. And, bankers do not like to lose money.)

Please note that the higher the discount rate, the lower the present value. This works in reverse for compounding. The higher the compounding rate, the higher the future value.

It is very tempting simply to add up all the annual payments for different loans and then compare them. But this is incorrect. The correct way to compare loans is to discount the future payments to their present value. The present value is their value in today's dollars. You must always compare loans in today's dollars.

Here is the brief methodology for comparing loans:

Method Rules

1. Write out the annual payment schedule, or amortization schedule, of each loan you are considering, including the amount in dollars that must be paid in each year along with the year in which it must be paid.
2. Choose the discount rate you will use.

3. Convert each annual payment to its present value by discounting it for the number of years in the future in which it must be paid.
4. Sum up all the annual present values for each financial option that you are considering.
5. Compare the present values of each alternative.

Points, Fees, and True Interest Cost (TIC)

It seems like it is totally impossible to borrow money without having to pay one or more fees or charges. That said, we need to know what impact they will have on our ADSPs that ratepayers must pay. This will lead us to the important concept of TIC, or, as the banks euphemistically refer to it, the Annual Percentage Rate (APR).

There are three types of fees or charges:

1. One-time only, upfront fees, such as the "points" the bank charges you when you take out your mortgage. This also includes other one-time charges such as attorney's fees, appraisal fees, and the like.
2. Annual fees that are expressed in flat dollar amounts. These are charges like a $25-per-month "servicing fee."
3. Annual fees that are expressed as a percentage of the outstanding principal balance. Some monitoring fees are expressed this way. For example, on a municipal bond, the bank, acting as bond trustee, might charge an annual monitoring fee of 1/8th of 1% (0.125%). (This actually happens, but it is a wonder why. Over time, the outstanding principal balance of the loan declines. So does the 0.125% fee. But the cost of monitoring [personnel, facilities, etc.] increases. So, you have the dire circumstance of declining income and rising expenses.)

Each of these types of charges is dealt with differently. Here is how they work:

1. One-time upfront fees are added to loan principal. So, when your bank charges you 2 points on your $200,000 mortgage, you actually have a $204,000 loan, not a $200,000 loan.
2. Annual fees are added to each annual payment. In this case, by adding a $25-a-month fee to your $1,113 monthly payment, it becomes a $1,138 monthly payment.
3. Annual fees that are expressed as a percentage of the outstanding principal balance are added to the interest rate. Here, your 6% loan with a 1/8th of 1% monitoring fee becomes a 6.125% interest rate.

This is all well and good, but what impact do these charges have on your monthly or annual payments? You should know this as a matter of personal finance, but all the more so if you are on a public board that is authorized to issue debt.

Finding the impact of one-time, upfront charges is very difficult without a financial calculator. And even then, it is not really easy.

Essentially, you use our favorite formula for calculating annual debt service payments, $ADSP = PV \times (i/(1 - (1/(1 + i)^n)))$, but we have to calculate backward to find i, the interest rate. In other words, the bank's adding points has the same impact on your loan as if they raised the interest rate. They could, of course, just raise the interest rate but then you might decline the loan. Also, they would get their money over 30 years. But with points upfront, they get the extra money today.

I do not see any reason to burden you with this gruesome calculation, so I will ask you to trust me. And, if you do not trust me, you can buy a financial calculator and do the calculations yourself.

If your bank charges you 2 points on your $200,000 mortgage and charges you a 5% rate of interest, here is what happens:

As we said before, you add the points to principal. So, you now have a $204,000 mortgage at 5% for, say, 15 years. In such case, your annual* payments would be $19,653.83. Now, if the bank did not charge you the 2 points (or you paid them in cash), you would have a $200,000 mortgage at 5% for 15 years. As such, your annual payment would be $19,268.46. So, what do the 2 points mean to your effective borrowing rate?

To calculate this effect, we take the ADSP on the $204,000 loan, but then take $200,000 as the loan principal, all at the 15-year term. This complicated effort reveals that your $200,000 mortgage with two points is the equivalent of having a 5.29% mortgage rate. When you finalize the paperwork for this loan, one of the little documents the bank will give you is a notice saying that, although the nominal rate of interest on your loan is 5%, your equivalent rate (because of the two points) is 5.29%. This is what the bankers call the *Annual Percentage Rate* (*APR*). Not too odious a euphemism. However, the rest of the finance industry calls this phenomenon the *True Interest Cost* (*TIC*), words that are much closer to the plain truth.

Annual flat fees are dealt with, as we said, by adding them to the annual payment. Let us look at a $100,000 loan for 20 years at a rate of 5%. The level annual payment on this loan would be $8,024.26. Adding $25 a month or $300 a year to this payment would make it $8,324.26. Now, what impact on your TIC does this $25 per month have?

This time, we take $8,324.26 as our annual payment, 20 years as our term, and $100,000 as our principal. Then we calculate a new interest rate, which will be our TIC. In this case, it is 5.44%. So in this case, our nominal rate of interest is 5%, but our TIC is 5.44%.

---

* Even though mortgages have monthly payments that are slightly different than simply multiplying them by 12 to get an annual payment, we are going to use a simple annual payment (like most public debt) for our examples here.

The last type of charge, an annual fee expressed as a percentage of the outstanding principal balance, is the easiest to deal with; all you do is add it to your interest rate. So, if your nominal interest rate is 5% and you are being charged a 1/8th of 1% (0.125%) monitoring fee, then your TIC is 5.125%.

Comparing loans or bonds, both with and without fees and charges, is essential in identifying the lowest cost alternative for ratepayers or the general public to pay. And identifying the lowest cost alternative is essential in providing the greatest environmental benefit to the largest possible number of people.

11

Hidden and Not-So-Hidden Cost Factors

Costs such as lawyers' fees are hardly hidden. Who could imagine a major financial transaction occurring without a battery of lawyers on each side? You cannot even have a mortgage settlement in most states without lawyers. On the other hand, however, ineligibility is a cost factor that can sneak up on you and bite you from behind.

There are many factors that inflate costs. Many come from the project itself. We, however, will consider only those that emanate from environmental finance programs.

In Chapter 9 we looked at coverage ratios. Coverage requirements force utilities to charge rates higher than those absolutely necessary to make annual debt service payments (ADSPs). These cost ratepayers money, but they cannot be considered hidden cost factors of the financing program itself. However, they are close.

We can break down the concept of hidden cost factors into two areas. First, there are those costs associated with the financing. Then there are some other miscellaneous costs that are extrinsic to the financing but, nonetheless, impact project costs. There are eight types of hidden costs that are directly related to the actual financing of projects.

Financing Costs

By financing costs we mean the costs of obtaining the funds themselves or of making the funding possible in the first place. These are not abstract concepts such as *term* and *rate*, but fees that must be considered in the overall project cost. As discussed in Chapter 10, these are one-time, upfront costs. As such, we calculate them into the project cost by adding them to the loan or bond principal. Below are the eight most common forms of financing costs.

Commitment Fees or Points

Commitment fees or points are commonly associated with bank loans as opposed to the issuance of municipal bonds. Although the nomenclature is by no means clear or universal, commitment fees are usually associated with commercial loans and sometimes even personal loans, as opposed to mortgages. Commitment fees are generally 1% to 2% of the loan principal amount and are paid upon acceptance of the bank's commitment to lend. The term *points*, on the other hand, most often refers to fees that are also 1% to 2% of loan principal and are paid at closing. On any given loan, there may be either, neither, or both, although the total of such fees seldom exceeds 2% of the loan principal.

As you have seen in Chapter 10, points are simply an additional payment to the lender that has the dual effect of increasing the banker's yield on the loan and also of increasing the borrower's ADSPs as well as his True Interest Cost (TIC). They simply constitute additional compensation to the lender and must be calculated into the real cost of the loan, whether by the annual payment method or the total payment method.

Commitment fees are, ostensibly, a form of compensation paid to the bank or lending institution for reserving funds from the time of commitment to the time of disbursement. In addition, they are also supposed to protect the bank against the possibility that the loan will not close, that is, that their already-reserved funds will not ultimately be taken up in a timely manner. We say "ostensibly" because, notwithstanding what they appear to be, or how anyone might choose to characterize them, commitment fees are simply a means of increasing the lender's yield. They, too, are additional compensation to the bank. Furthermore, if their true purpose were to protect the bank against the borrower's failure to close the loan, then the commitment fee would be refundable to the borrower at closing, would it not? But, of course, it is not.

In addition, if commitment fees were actually related to the bank's cost of reserving funds, then they would have two characteristics. First, the amount of the fee would relate directly to the bank's cost of funds. This means that if the bank's cost of funds were 12% a year, then the commitment fee should be 1% per month, no? If the bank's cost of funds were 6% per year, then it should charge 0.5% per month for its commitments. But that is not the way it works. As we said before, banks charge 1% to 2% regardless of whether the closing is in 1 month or 6 months and whether their cost of funds is 10% or 0%.

Second, again if the bank's fees were related to its cost of funds, then the fees should be higher the longer the period between commitment and settlement. This would mean, again, that the fees would be something like 1% or 0.5% *per month*. But, as you know, they are not.

Financial Advisory Fees

A good financial advisor (FA) is the most valuable member of the team of experts who advise your system on financing its projects. The FA works for the utility or system. He/she should not be confused with an underwriter who is independent and who should be willing to actually buy your bonds. The FA does not buy bonds. The FA guides the sale.

Your FA should be your closest advisor on each transaction. FAs should actually drive the deals. In terms of credentials, they should have done many similar deals in your state. They should be familiar with the market for bonds in your state, as well as the terms and conditions of any environmental finance programs. A good FA can help you choose a bond counsel and advise you on almost every aspect of the deal. Although not a lawyer, and therefore unable to issue a bond opinion, a good FA will know bond law from A to Z.

Back in the day, FAs were paid a fee based exclusively on the size of the deal. Things, however, have changed. FAs need to earn, I would say, a minimum of $50,000 on each deal. Remember: It is not just the person you are dealing with; rather, it is he/she and his/her support staff and his/her office rent and other expenses. So, nowadays, I think $50,000 is a minimum.

This raises a ticklish question. If you express the FA's fee as a percentage of bond principal, then on deals in the $1-million range, the FA's $50,000 fee would be a breathtaking 5% of the deal! Numbers of this size just do not sit well with ratepayers or the general public. So, instead, you and the FA might agree on a split fee. Maybe you could pay the FA $2,500 a month for a year, or $5,000 a month for six months, which would equal $30,000. Then you could pay the FA an additional 2%, or $20,000, more at the closing of the bond issue.

As the size of your bond issue increases, so too will the fee that the FA wants. When you get up into the $8- to $10-million range, the FA should expect to be paid at least $200,000. In my experience, if you have a good, smart FA with whom you are both comfortable and compatible, the fees you pay are definitely going to be worth every penny.

Counsel Fees

Sometimes there are so many lawyers involved in a deal that it makes you laugh. Unless, of course, it is you who must pay all of them. Some of the different kinds of lawyers involved in large financial transactions include house counsel, outside counsel, lender's counsel, underwriter's counsel, and bond counsel.

The good news is that, depending on whether you are financing with a bond or a bank loan, you will not need all five kinds of lawyers. The bad news is that you will probably need four of them:

- House counsel is the lawyer who works for your system (utility). He is on salary. You do not need to pay him anything extra.
- Outside counsel you may, or may not, need. If your system undertakes a major project—one involving, say, several million dollars that needs to be financed over an extended term—only once every 5 to 10 years, then it is unlikely that your house counsel has the knowledge or the skills to guide your system through a financing of this magnitude. So you will need to engage outside counsel. You can do this by competitive bid. You can issue a Request For Proposal (RFP). They will be paid at closing out of the bond proceeds. They will

charge hourly, so you will need to estimate their fees before closing. If you can get away with paying a skilled financial lawyer less than $500 an hour, you are lucky indeed. You will probably have to pay more. If the lawyer quotes a fee substantially below $500 an hour, then thoroughly examine his credentials. Call some of his other clients and check him out thoroughly.

- Lender's counsel represents the bank, if you are using one.
- Underwriter's counsel represents the firm that is buying your bonds, if that is the financing method you choose. They both charge fees you would expect from sophisticated financial lawyers: $500 an hour and up.
- Bond counsel is the lawyer who issues an independent opinion that the bonds you have issued are valid debt of your system and that they are due and payable according to their terms. Most importantly, if you are issuing a tax-exempt municipal bond, then this lawyer will issue an opinion that the interest on your bonds is exempt from federal (and most probably state) income taxation. This is the key to a successful municipal bond issue. And, if the bond counsel is wrong, for any reason, and the Internal Revenue Service determines that the interest on your bonds *is* taxable, then the bond trustee will sue the stuffing out of that bond counsel on behalf of all the bondholders. You can bet that bond counsel will charge a hefty fee. Sometimes this fee is based on hours; other times, it is simply a negotiated fee. But you can certainly bet that, whether flat or hourly, the bond counsel's fee will be very substantial.

Servicing Fees

Servicing fees occur in both bond and loan transactions. They cover the administrative costs of collecting your system's loan or bond payments, and, in the case of a bond offering, paying those funds to the possibly thousands of bondholders who bought them.

In the case of a bond offering, the servicing fees are usually called *trustee's fees*. They are so called because each bond issue has a bank that acts as a trustee who handles your system's payments.

As you know, the underwriter that buys your bonds does not keep them. The underwriter resells the bonds to its wholesale customers, such as mutual funds that deal in bonds, or directly at retail to everyone's "favorite grandmother." So, literally thousands of "grandmothers" may own your system's bonds.

The bank trustee, then, has an important job. The trustee is legally a fiduciary and has strict legal responsibility to each of the thousands of "grandmothers" who own your bonds. Because there is legal liability involved, and because there is considerable work involved in keeping track of the thousands of "grandmothers" and how much each is owed, trustees' fees are generally higher than servicing fees on a simple bank loan where all the servicer has to do is credit your account and debit the bank's account with your payment.

Trustees' fees are generally expressed as a percentage of outstanding principal on the bond. Bank loan servicing fees are generally a flat number. Flat fees make more sense than fees based on principal balances. The salaries of the actual people who perform the servicer's/trustee's duties go up over time. Over the same period, the flat fee remains flat and, worse, the principal balance-based fee actually declines. Rising expenses and declining income is not a recipe for a successful business.

Trustees' fees are generally in the range of a tenth of 1% (0.1%) to a quarter of 1% (0.25%). As noted before, servicing fees are lower and they are flat. They are generally nonnegotiable.

Underwriting/Placement Fees

Underwriting fees, generally called the underwriter's discount, and placement fees are the ways that investment bankers get paid for buying your bond. And, as noted above, the notion that they actually "buy" your bonds is somewhat fatuous. Yes, they actually wire funds to your account, and, yes, they take physical possession of the bonds (unless, of course, the transaction is totally electronic), but they only "own" the bonds for a split second. The instant the bonds are transferred to the underwriter's account, he distributes them to the people to whom he has pre-sold them. Likewise, he credits their individual accounts for the amount of bonds they purchased and debits his own account with their funds. The underwriter thus reimburses himself for the money he paid you plus, of course, a profit.

Back in the day, the relationships between borrowers and their underwriters and placement agents were often quite clubby. Utility systems and even local governments just used to hire, or engage, underwriters or placement agents. No competitive bidding, just relationships. And the fees that the underwriters and placement agents charged were "negotiated" with their friends at the utility system or local government. And, in the spirit of the old Latin maxim, *manus manum lavat* (one hand washes the other), if the board positions were elected offices, you often found underwriters and placement agents making substantial political contributions to the incumbents. Very nice arrangement, while it lasted.

Furthermore, if a bond issue required a public referendum, underwriters and placement agents often made hefty contributions to publicity efforts promoting the bond issue. Not so much anymore. The Securities and Exchange Commission (SEC), and more recently the Finance Industry Regulatory Agency (FINRA), have come down hard on these practices.

Underwriters/placement agents are usually paid a percentage of bond principal. These fees can be in the form of a sliding scale according to maturity.

In other words, the longer the maturity, the higher will be the underwriter's percentage fee.

In addition, much like the FA's fees, the underwriter's discount will be larger than 1% if the bond issue is small, such as under $10 million. But if the issue is large, such as over $100 million, the underwriter's discount will likely be less than 0.5%.

Credit Enhancement Fees

Before the sub-prime mortgage crisis, there were 15 financial guaranty insurance companies, or as they are more commonly called, municipal bond insurance companies, in the United States. In 2005, bond insurers guaranteed almost 60% of the about $400 billion of municipal bonds issued that year. In 2011, there was only one company left, which insured only about 5% of the market.

Since 2011, another company has entered the market, Build America Mutual Assurance Corporation (BAM). In addition, the surviving company, Assured Guaranty, spun off a municipals-only subsidiary in 2013, called the Municipal Assurance Corporation (MAC), which is rated AA– by Standard & Poor's. BAM is rated AA.

Regardless of their ratings, because of the rocky history of bond insurers, their insured bonds sometimes do not trade as well as an uninsured bond with the same rating. In other words, if City A, with a AA credit rating issues a bond, and City B, with only a BBB rating, but which buys insurance from BAM, issues an identical bond, you will often see City A's bond trading at a better rate than City B's bond.

So, the appropriate way to deal with bond insurance is to confer with your FA to determine just how (at what rates) the two bond insurers are trading. Ask the FA where your bonds would trade without insurance and request a quote from the two insurers. Then sit down with your FA to see if you will have enough savings to justify the purchase of the bond insurance.

In general, the rule is that the insurance premium should eat up just under half of the spread between what your bonds would trade at with the insurance versus without the insurance. In other words, if your bonds would sell at 5% without the insurance and 4.5% with the insurance, then the premium should cost you about 0.2%, which means your all-in rate with the insurance would be 4.7%. So you would save the difference between 5% and 4.7%, or 0.3%.

Back in the day, banks used to compete with the bond insurers by offering Letters of Credit (LoCs) to guaranty bonds. The collapse of the municipal bond market has scared off most of them, especially from long-term guaranties. But, if you do encounter a bank willing to issue an LoC to back your bonds, treat them just as you would a bond insurer.

Rating Agency Fees

This is the simplest cost element to understand. If your system intends to issue bonds and your FA believes your system has the credit strength to obtain an investment-grade rating on the bonds, then he undoubtedly will recommend that the system apply for one. The fee is the cost of obtaining the rating. This fee is paid directly to the rating agency at the closing, when they formally issue their credit rating on your bond issue.

The rating agency fee is a flat fee and is based on the complexity of the review, which the rating agency staff conducts of your system's financial statements, management procedures, and operating history. Fees range from a few thousand dollars for a smaller system with simple accounting procedures and management history, to several tens of thousands of dollars for larger, more complicated systems.

The rating agencies will agree to the amount of their fees in advance. It is part of the FA's job to contact the rating agencies on your system's behalf and negotiate the rating fee. If you have a decent-sized bond issue (e.g., over $10 million), your FA will probably recommend that you get ratings from more than one agency.

The FA will also be able to guide your system's staff through the issues the rating agencies will want to know about.

Rating agency staff members are very professional and their reviews are quite rigorous. It is a fair thing to say that if you apply for a rating, the fee charged by the rating agency will be the least of your worries.

Miscellaneous Imposed Costs

In every financing, there will always be some miscellaneous costs but they should not involve large amounts of money. They should also be easily able to identify in advance. Among the more common types of miscellaneous costs are

- *The cost of printing bond documents.* The document that legally offers the bonds for sale is called the "Official Statement," or "OS." Investment bankers—even in this electronic world—have some OSs printed by a professional printer. It costs very little—maybe $1,000 to $2,000.
- *Surveys.* Whenever a piece of real estate is involved in a transaction, the buyer (which would be the utility system) usually requires that a survey be done by a professional surveyor just to make absolutely certain of the exact boundaries of the property. The cost of the survey can run from a few hundred dollars for a relatively small property to several thousand dollars for a large, complicated parcel of land.
- *Title insurance.* Title insurance performs two functions in a transaction involving the financing of a purchase of real estate. First, it protects the owner from defects in title. It indemnifies the owner if

another party can prove that the owner's title is not valid or defective. A title policy can also protect a mortgagee (bank) from an invalid or unenforceable mortgage. Such a policy would indemnify the bank for any losses it sustained because of either of those two causes.

- *Appraisals.* Again, when real property is involved in a transaction, appraisals are almost always required. Professional appraisers have the initials "M.A.I." after their names, which stand for Member of the Appraisal Institute. You should only do business with these professionals. The cost of an appraisal can run several thousands of dollars for a sizeable piece of property.

These miscellaneous imposed costs only amount to a few thousand dollars; but for smaller transactions, this can be significant.

Nonfinancial Additional Cost Factors

There are two additional nonfinancial factors that we must consider: *delay* and *ineligibility*. The concept of ineligibility has some unexpected twists, such as "Buy American" requirements and Davis–Bacon, or prevailing labor wage, requirements.

Delay

Delay can be a serious matter and can sneak up on you. Take the Clean Water State Revolving Fund (CWSRF), for example. First of all, from the time your consulting engineers make their presentation calling for a new capital project until the time you are ready to finance the project will probably take six months to one year, just because there are so many pieces that must be put together and so many actors/advisors to hire, all of which needs to be done by competitive bidding and which can be quite time consuming. Project budgets must take into account inflation and the time value of money, as was outlined in Chapter 10. For example, at the beginning of "Year 1," an infrastructure project is estimated to cost $1,000,000. The utility decides to implement the project but is forced to wait three years due to bureaucratic delays. In three years, the "cost" is no longer $1,000,000; the project cost must be compounded to a future value. Assuming a 3% rate of inflation, the new project cost will be almost $93,000 more than the original estimate: future value of project cost = $1,000,000 × $(1.03)^3$ = $1,092,727. This is not an insignificant increase. It is more than 9%. If there is a 1:1 correlation between your ADSP and your system's rates, then the ratepayers will be faced with an additional (over and above the cost of the project itself) 9% cost increase.

In some states there is an annual deadline for applications for the coming year. If you miss the deadline, you have to wait another year to apply.

Most states require Environmental Appraisals, which, if some significant issues are identified, will then require a full Environmental Impact Statement (EIS). If this happens, you will spend at least another year writing the specs, going out with an RFP, selecting the independent engineer to do the statement, and then waiting for him to get the work done and get it approved by the state.

Every time a year goes by, costs go up by about 3%. And, in the past few years, construction costs have escalated much faster than has the Consumer Price Index (CPI), which is the usual measure of inflation. So, delay may well cost you far more than the simple rate of inflation.

Ineligibility

Ineligibility means that a lender's particular program to fund the system's project will not permit certain categories of costs to be included. Ineligible costs do not directly add to project costs, but do create a second project, from a financial perspective, and must be financed in other ways.

There are two categories of ineligibility: quantitative ineligibility and qualitative ineligibility.

- *Quantitative ineligibility:* Refers to a percentage of the total cost, which a lender will not finance. Quantitative ineligibility is a concept that you run into with banks, not the bond market.

As you know, before the sub-prime real estate hysteria, it was virtually impossible to get a bank to give you a 100% mortgage on a new home. A large part of this policy relates to liquidity. If a bank gave you a 100% mortgage, and you defaulted, and the market had slumped by 10%, the bank would be out a considerable amount of money. So, there is a principle here: The higher the loan-to-value ratio, the less liquid (and, therefore, the riskier) is their loan.

Banks that rely on assets as collateral need liquidity. The only way they can get liquidity is to lend less on the assets. That way, if the borrower defaults, they can get their money back quickly. For example, let us say that the bank will lend a construction company $50,000 on a piece of heavy machinery that costs $100,000. Let us say that the construction company defaults on the loan after a few months and the bank repossesses the piece of equipment. The next thing the bank needs to do is to sell it. So, because they only have $50,000 invested in the equipment, and because the equipment is less than a year old, they can probably sell it for $75,000 to $80,000, which means that the bank will get back every penny of its $50,000 loan.

Here are some of the categories of commercial loans where there is quantitative ineligibility:

• Raw land	50%
• Buildings, machinery and equipment	75%–80%
• Inventory or other disposable assets	60%–70%

One of the most important considerations for quantitative ineligibility is that you must finance the ineligible cost yourself. You cannot go to Bank A and get a 50% loan on a piece of raw land and then go to Bank B and get another loan for the 50% balance.* No, when Bank A makes you the loan, they will put a lien on the property in the county clerk's office. Before Bank B gives you a loan, they will search the county clerk's records to see if there are any prior liens on the property. When they find Bank A's lien, Bank B simply tears up your loan application.

- *Qualitative ineligibility*: Refers to whole categories of costs, which will not be considered for financing. Here are some examples taken from the requirements of the CWSRF:
  - Land. Under the CWSRF, land is an ineligible cost. If you have a $2,000,000 project that involves purchasing, say, $200,000 of land, then as far as the CWSRF is concerned, you have a $1,800,000 project because the $200,000 land cost is ineligible.
  - Goods and services that are not competitively bid. The CWSRF requires that all goods and services financed be competitively bid. There are certain cases, however, as with consulting engineers, where a water or wastewater system will have a long-term, ongoing, very satisfactory relationship with a particular consulting engineering firm. In such a case, they are not going to put the engineering services related to their project up for competitive bid. These engineering fees may be up to 6% of the total project cost. No matter—if the services are not bid, the cost is ineligible.
  - Non-Davis–Bacon wages. As explained in Chapter 1, so-called Davis–Bacon wages are simply the prevailing wages in a given area, such as a county. These rates are published. Prevailing wage rates have come to be associated with union wage rates. Where there are strong construction unions, the "prevailing" wages tend to be the union wage rate. But this is not necessarily so. In the United States, 86% of construction labor is non-union. In areas with little union activity, there are still prevailing wages, which are defined as the hourly wage, usual benefits, and overtime, that are paid to the majority of workers, laborers, and mechanics in a particular area. These "prevailing" wage rates are ascertained by either state departments of labor or an equivalent agency.

    To the extent that there are nonprevailing wages involved in any project, those costs are ineligible for funding under the CWSRF. In certain states, the payment of nonprevailing wages

* The exception to this, of course, is to get the seller of the property to take back a 50% subordinated note, or second mortgage. However, it would be an imprudent seller who would give you the full 50%; maybe 25% to 30% would be more like it.

might make the entire project ineligible for any CWSRF funding whatsoever.

- Buy American. The American Recovery and Reinvestment Act of 2009 (ARRA) inserted a "Buy American" provision into the CWSRF requirements. The requirement expired with the ARRA money, but the US Senate has been trying to reinsert it ever since. This provision may not cause any problems in many cases; but in some cases, where foreign goods are of comparable quality but significantly cheaper, or where there is a definitely superior foreign technology, they can cause problems. Because this requirement was completely unexpected in 2009, it caused quite a bit of chaos. The state CWSRFs are not anxious to see the "Buy American" requirement reinstated.

A final note about qualitative ineligibility. As is the case with quantitative ineligibility, those ineligible parts of a project may be extremely difficult to otherwise finance. Take the case of the $200,000 of land in the above example. The utility could not simply borrow the $1,800,000 from the CWSRF and then issue a tax-exempt municipal bond for the $200,000 land cost. They would get eaten alive by the cost of issuing such a small bond.

The utility might try to get a bank loan, but banks do not like to lend money for long periods of time. The utility might be able to get an 80% loan for 20 years from some lender, but they would be lucky. If anything went wrong, what would the bank do with the land that happened to be sitting under a sewage treatment plant? They could, of course, foreclose; but if the utility was not making its mortgage payments anyway, they would undoubtedly be going through some major financial difficulties, which means that the bank would not get them to pay in full. And, if the bank did foreclose, to whom could they sell a piece of land sitting under a sewage treatment plant?

Banks like to make 80% commercial mortgage loans on nice office buildings or nice strip malls. Sewage treatment plants are not going to be on their favorites list, for sure.

In summary, there are many hidden and not-so-hidden costs that people need to be aware of so that they can take appropriate actions to minimize them. We must remember that the lowest possible cost is our goal, so that we can bring the greatest environmental good to the largest number of people.

12

Impact of Term on Annual Debt Service Payment

We must begin the analysis by saying that the term of a loan has an enormous impact on its Annual Debt Service Payment (ADSP). This is true whether you choose the level payment method or the level principal payment method.

You may recall that, in Chapter 9, we briefly discussed these two most widely used methods of repaying the principal of a loan: the level payment method and the level principal payment method. These concepts take on great importance when discussing the impact of term on ADSPs. As you will see, the impact of term on annual debt service makes the decision of whether to choose a loan with the level payment method or the level principal payment method a critical judgment. For this reason, we will discuss the impact of term on annual debt service separately for each type of payment method. Then we will compare the two types of loan payment methods in terms of their respective impacts on ADSP.

Level Principal Payment Loans

If the term of a loan is only 1 year, then 100% of the principal is due and payable at the end of that year. If the term of a loan is 2 years, then 100% of the principal must be repaid by the end of the second year. (Note that we say "by" the end of the second year, not "at" the end of the second year. If the latter were true, the loan would be a balloon loan.)

Let us assume that we are dealing with a level principal payment loan, so that 50% of the principal balance is due and payable at the end of each year. Then, for a $100,000 loan at an interest rate of 7%, the ADSP at the end of the first year would be $57,000, consisting of $50,000 (or 50%) of principal and $7,000 of interest, representing 7% on the $100,000 of principal that was outstanding during the first year of the loan.

The second ADSP, however, which would be due by the end of year 2, would only be $53,500. This payment is composed of $50,000 of principal, which is the remaining principal outstanding, and $3,500 of interest, which is 7% of the $50,000 of principal that was outstanding during the second year of the loan.

Therefore, the ADSP at the end of year 1 would be $57,000, and the ADSP at the end of year 2 would be $53,500. The difference between these two payments is only $3,500, which is about 7%. Even if the amount of the loan were increased from $100,000 to $1,000,000, the difference in the two ADSPs would be only $35,000, which is, of course, still only 7%.

A 7% difference in ADSP cannot be considered significant. After all, if you consider all the cost factors that go into a water bill, or a sewer bill, or a solid waste bill, then ADSPs only make up a fraction of the total annual service charges to system users. For example, if the operation and maintenance expense for a water or wastewater system were only equal to annual debt service payments, then operation and maintenance expenses would be 50% of total system user charges and ADSPs would be 50% of total system user charges. In such a case, the 7% difference in the two ADSPs in our example above would only necessitate a 3.5% difference in total system user charges, as 7% of 50% is equal to 3.5% of 100%.

In our example, the first year's ADSP is $57,000. If the operation and maintenance expenses that year were also $57,000, then, theoretically, the total system user charge for the year would be $114,000. Assuming the operation and maintenance expenses stayed the same in year 2, the second ADSP would be only $53,500. Total system user charges for the year would be reduced to $113,500, which is about a 3.5% decrease. A 3.5% difference in total system user charges from one year to the next is not really significant—especially when it is a *de*crease.

However, when we start talking about loans with terms of more than two years, the story begins to change radically.

Let us use the same $100,000 loan and the same 7% interest rate, but this time use a 20-year term. In doing so, we can examine the difference between the first ADSP and the last ADSP.

Assuming, once again, that we are dealing with a level principal payment loan, the amount of principal that would be repaid in every year would be one-twentieth (1/20) of $100,000, or $5,000. For the first year, the outstanding principal balance would be the original principal balance, or $100,000. The interest on $100,000 for 1 year at a rate of 7% would be $7,000. Because the principal payment at the end of the first year would be $5,000 and the interest payment would be $7,000, the total ADSP for year 1 would be $12,000.

For year 20, the outstanding principal balance would only be $5,000, because the other $95,000 had been repaid in the preceding 19 years. The interest, at a rate of 7%, on the outstanding principal balance of $5,000 would be $350. Therefore, the total ADSP for year 20 would be $350 of interest and $5,000 of principal, or $5,350.

Now compare the first year's ADSP of $12,000 with the ADSP in year 20 of $5,350. Here, the difference in the two ADSPs is $6,650, or a 55% decrease! We can be certain that the ratepayers in year 20 would be pleased with this circumstance, but what about the ratepayers in year 1 who are getting the benefit of the same project?

In Chapter 9 we discussed level payment loans and level principal payment loans. We observed that one of the key points in favor of level payment loans involved the concept of fairness. However, we did not go into the matter in detail. Let us illustrate this point now with an example.

As you will recall, a level payment loan means that each ADSP is equal to any other ADSP. If the service life of, say, a new treatment facility is 20 years, then using our example above of a $100,000 loan at an interest rate of 7%, the ADSP under the level payment method would be $9,439. This means that the ADSP would be $9,439 in each of the 20 years the loan was outstanding. In such a case, the total debt service would be twenty times $9,439 (20 × $9,439), or $188,780.

A ratepayer in year 1 would pay his share of the $9,439, which is one-twentieth of the total project cost. Assuming that the project—say a new treatment facility—had a service life of 20 years, then that ratepayer would also enjoy one-twentieth, or 5%, of the benefit of the project. In other words, in year 1, he would pay 5% of the cost of the project and he would receive 5% of the benefit of the project. Fair enough.

Now, what if the project loan were a level principal payment loan instead of a level payment loan? Table 12.1 sets forth the annual principal payments, the annual interest payments, and the annual debt service payments on a 20-year, level principal payment loan of $100,000 at an interest rate of 7%*.

As you can see, with the level principal payment loan, the ratepayer in year 1 pays 6.9% of the project cost. The problem is that, because the service life of the project is only 20 years, he only gets the benefit of one-twentieth, or 5%, of the project. So, he winds up paying for 6.9% of the project but getting the benefit of only 5% of the project.

By the same token, look now at the ratepayer in year 20. He pays his proportionate share of the ADSP of just $5,350, which is only 3.1% of total project cost. So, the year-20 ratepayer only pays for 3.1% of the project but winds up getting the benefit of a full 5% of the project.

Now, 6.9%, 5%, and 3.1% look like pretty small numbers; but if you want to characterize the ADSPs in years 1 and 20, you can see that the difference between 6.9 and 3.1 is 123%. This means that the ADSP in year 1 is 123% *higher* than the twentieth ADSP.

The differences between 3.1% and 6.9% are not much, but system ratepayers will certainly want to know why they are being charged 123% more when they are only getting the same benefit out of the project!

Now, the theme of this chapter concerns the effect of rate on term. Let us look, without going into all the details, at the same two loans if they had 40-year terms instead of 20-year terms.

* Please note that the last column in Table 12.1 is captioned "% of Total Debt Service." The percentages listed on each line represent the ADSP for that year, divided by the total debt service.

TABLE 12.1

Year	Annual Principal Payment	Annual Interest Payment	Annual Debt Service Payment	% of Total Debt Service
1	$5,000	$7,000	$12,000	6.90
2	$5,000	$6,660	$11,650	6.70
3	$5,000	$6,300	$11,300	6.60
4	$5,000	$5,960	$10,950	6.30
6	$5,000	$6,600	$10,600	6.10
6	$5,000	$6,250	$10,260	6.90
7	$5,000	$4,900	$9,900	5.70
8	$5,000	$4,650	$9,550	6.60
9	$5,000	$4,200	$9,200	6.30
10	$5,000	$3,850	$8,850	6.10
11	$5,000	$3,500	$8,500	4.90
12	$5,000	$3,160	$8,150	4.70
13	$5,000	$2,800	$7,800	4.60
14	$5,000	$2,450	$7,450	4.30
16	$5,000	$2,100	$7,100	4.10
16	$5,000	$1,750	$6,750	3.90
17	$5,000	$1,400	$6,400	3.70
18	$5,000	$1,050	$6,050	3.50
19	$5,000	$700	$5,700	3.30
20	$5,000	$350	$5 350	3.10
	$100,000	**$73,500**	**$173,600**	**100**

On the level principal loan, the annual principal payment on a 40-year, $100,000 loan would be one-fortieth (1/40), 2.5%, or $2,500. The interest payment would be the same in year 1 as for the 20-year loan (because the full $100,000 is outstanding in the first year), or $7,000. So, the ADSP in year 1 for the level principal payment loan would be $9,500. For the fortieth year, only $2,500 would remain outstanding. The interest on this amount, at the 7% rate, would be $175. As a consequence, the fortieth year ADSP would be $2,675.

The difference between the ADSP in year 1 ($9,500) can be characterized as being 258% *higher* than the ADSP in year 40 ($2,675). How do you think your ratepayers would react to those numbers?

Now that we have reviewed the impact of term on the ADSP of a level principal payment loan, we will move on to the consideration of the impact of term on the ADSPs of a level payment loan.

To put the concept of the power of term into full perspective when dealing with level principal payment loans, the table below sets forth the first years' annual payments on a $100 loan at several different interest rates over several different terms:

	Interest Rates				
Term	0%	5%	10%	15%	20%
1 year	$100	$105	$110	$115	$120
2 years	$50	$55	$60	$65	$70
3 years	$33	$38	$43	$48	$53
4 years	$25	$30	$35	$40	$45
5 years	$20	$25	$30	$35	$40
10 years	$10	$15	$20	$25	$30
20 years	$5	$10	$15	$20	$25
30 years	$3	$8	$13	$18	$23
40 years	$2.50	$7.50	$12.50	$17.50	$22.50

Please note the hugely significant impact that term has on the ADSPs for these loans.

And now we turn to the level payment method.

Level Payment Method

The impact of term on ADSPs is as significant when the level payment method is used as when the level principal payment method is used. With the level payment method, however, there is no problem with fairness. There is no problem with an adverse characterization of the numbers.

	Interest Rates				
Term	0%	5%	10%	15%	20%
1 year	$100	$105	$110	$115	$120
2 years	$50	$54	$58	$62	$65
3 years	$33	$37	$40	$44	$47
4 years	$25	$28	$32	$35	$39
5 years	$20	$23	$26	$30	$33
10 years	$10	$13	$16	$20	$24
20 years	$5	$8	$12	$16	$21
30 years	$3	$7	$11	$15	$20
40 years	$2.50	$6	$10	$15	$20

Again, you can see the enormous impact of term on ADSPs. The payment on a 10-year loan at 5% is $13, whereas it is $8 on a 20-year loan. This means that the 10-year loan costs ratepayers 63% more than the 20-year loan. And, of course, if the service lives of the assets being financed were 20 years, why would any utility favor the shorter-term loan?

Now, as we have said several times so far in this book, *term* is one of the most powerful concepts for lowering the cost of environmental improvement projects. And, as we know, the less expensive such projects are, the more will get done.

I hope you will recall our doctor in Chautauqua County, New York. He chose a 0-term loan; that is, he paid cash . . . $60,000 in cash. Let us assume that us mortals might need a geothermal project of one-tenth the size, so a $6,000 project. So, using the same numbers, our monthly savings would be $100. Now, being mortals, most of us do not have $6,000 in cash to drop for a geothermal system. But if our county had a Property Assessed Clean Energy (PACE)-type program where we could finance the project for 30 years, then our monthly payment would be $32.50. So our savings would be $100 and our cost $32.50, for a net savings, or profit, of $67.50 a month. As such, it is likely that many more people would undertake such projects . . . all because of the savings . . . all because of the power of term.

A few years ago, when California was beginning to organize its PACE program, I met an air quality officer in the Central Valley. She told me she had just taken out a second mortgage to install insulation in her house and to replace all the doors and windows. She said that her bank gave her a 7% loan for seven years. Assuming she borrowed $10,000, her monthly payment would have been $151.

Now, if her county had organized a PACE program, she could have gotten a 20-year term (without the second mortgage) loan. Even assuming the same 7% rate of interest, her monthly payment would have been $78!

Recall that the first principle of environmental finance is to create the greatest amount of environmental benefit for the largest number of people at the lowest cost. The second principle is that the more we can drive down the cost of environmental improvements, the more will get done. This is a golden example of these two principles. More people are likely to insulate their homes at a cost of $78, than at $151. And that is our goal!

Term is one of the most powerful mechanisms for driving down the cost of environmental improvement projects. And, the lower the cost, the more such projects will get done.

13

Grants and Affordability

Grants

In any book about environmental finance, one would think that a considerable amount of ink would be devoted to the subject of grants. After all, grants are certainly the most popular method of financing anything... anything, that is, related to the government. Of course they are popular! They are free!

Notwithstanding their popularity, the section in this book on grants will be quite brief. That is because in my quarter of a century observing governmental environmental finance programs, I have identified only four circumstances that I believe are legitimate uses for grants. Other uses are just political; they do not seem to serve any other purpose.

The four uses of grants, which I deem legitimate, are as follows:

1. *Paying for environmental services that are not affordable, either to individuals or communities.* Remember Jack and Diane in Chapter 1. They desperately needed to buy a new wood stove to keep their home warm and their baby daughter healthy. What form of government assistance was available to them? A tax credit, which they were too poor to be able to use. What did they need? A grant. Either a 100% grant, or something very close to 100%. That is an example for individuals.

   A good example of a sound community grant is in the case of the US Department of Agriculture (USDA) water and wastewater program. Here, when they see that a project costs a significant share of median household income and will make rates higher than in surrounding districts, then the USDA uses grant money to buy down the cost of the project to levels where it will be affordable to the average ratepayer.
2. *Inducing people or businesses to make environmental improvements that they are not legally required to do.* When the American Recovery and Reinvestment Act (ARRA) was passed in 2009 to help us get out of the sub-prime mortgage disaster, *renewables* was just beginning to be a major buzzword. The ARRA included substantial grant funds for

those who would reduce their carbon footprint by installing solar panels. No one was required to install any kind of renewable or energy-saving device. So, the government paid them to do so.

Cover crops are another good example. After crops are harvested, a large amount of nitrogen remains in the soil. Over the winter with snows and rain on the now-bare soil, much of this nitrogen migrates to the nearest water body where it pollutes the water. Cover crops are planted after the main crop is harvested. Their sole purpose is to sop up nitrogen left over in the soil to prevent it from polluting any streams or ponds. In the United States, we pay (i.e., give grants to) farmers who will plant cover crops. These grants come out of the "farm bills" that the Congress passes to maintain its elaborate scheme of subsidies for agriculture. In Germany, they have a cover crop program too. It is also a grant program, but in Germany the grants come from a special tax that is collected each year.

3. *Creating or commercializing new environmental technologies.* Until most recently, it was not cost effective to install solar panels. Even with generous government financing programs, the monthly cost of the panels exceeded the savings in electricity from the panels. In order to jump-start the solar industry, the US Department of Energy started a grant (i.e., subsidy program) for solar panel manufacturers. The Chinese did the same. However, the Chinese subsidies were so large that the Chinese manufacturers were able to sell their solar panels for below the cost of production. The Chinese government created subsidies so that their manufacturers could export all over the world, creating thousands of good manufacturing jobs in China in the process. The United States, European Union, and China are now embroiled in a trade war over this issue. Regardless of the trade issue, using grants to jump-start new, needed, environmental technologies is a very good idea.
4. *Environmental education.* Giving out smaller grants is a good way to get community groups to take an avid interest in the environment. This is especially so for poorer groups that cannot readily raise funds for projects. Maryland has its Chesapeake and Atlantic Coastal Bays Trust Fund, which most natives call the "green fund." This fund is relatively small but it gives out grants to community groups for such activities as streambed restoration, tree plantings, and the like. Educating people about the value of the environment and the steps necessary to protect it is certainly a valid expenditure of public funds. And, in this case, it really needs to be grant funds.

A final word about the *use* of grant funds. In general, because of inflation and the fact that the value of money shrinks over time, grants should be used upfront. They should be used to buy down the cost of a project. Conversely, the only times you should make annual grants are when you have a grant fund, like Maryland's "green fund." Here, the key consideration is that the grants go to different groups every year. The grants are competed out.

Another legitimate use of annual grant funds is the aforementioned cover crops. Here, the grants go to the same people each year, participating farmers, but the activity is annual. It is not like buying or building a facility. With cover crops, you are more or less paying for annual labor, not one-time bricks and mortar.

An illustrative example of why one-time, upfront grants are preferable to annual grants when facilities are being built occurred in Ukraine in 2005. At that time, the State Committee for Housing and Communal Services* was considering the use of state appropriations as grants to reduce project costs to local governments. At the time, commercial bank lending rates in the Ukraine were 20%. Local mayors who needed projects went to the central government and complained about the high bank interest rates. The government gave in to their demands and created a 5-year, 0% loan program to the local governments. So, the game plan for the program was that the local mayors would borrow money from their local banks and pay back the principal in five equal annual installments (the level principal payment method). The government would then appropriate funds and pay the interest on these local government loans. Table 13.1 illustrates what this looked like.

It is clear from this analysis that the present value of the State Committee's indirect subsidy of 50.37 equals slightly more than 50% of the entire project cost of UAH100.†

TABLE 13.1

Annual Grant

	Year						
	0	**1**	**2**	**3**	**4**	**5**	**Total**
Total Financing Required	100						
Upfront Grant by State Committee	0						
Outstanding Balance	100	80	60	40	20	0	n/a
Principal Paid by Local Gov't	0	20	20	20	20	20	100
Interest Paid by State Committee	0	20	16	12	8	4	60
Present Value (PV) of Subsidized Interest Payment	0	18.52	13.72	9.53	5.88	2.72	50.37

Note: The inflation rate in the Ukraine at this time was 8%, so that is what was used as the discount rate.

* Communal Services were water and wastewater.

† UAH are Ukrainian hryvnia, the country's official currency.

TABLE 13.2
Upfront Grant

	Year						
	0	**1**	**2**	**3**	**4**	**5**	**Total**
Total Financing Required	100						
Subsidy by State Committee	50						
Outstanding Balance	50	40	30	20	10	0	n/a
Principal Paid by Local Gov't	0	10	10	10	10	10	50
Interest Paid by Local Gov't	0	10	8	6	4	2	30
Total Paid by Local Gov't	0	20	18	16	14	12	80

Now let us see what the effect would be if that same amount of money (UAH50) were used upfront to reduce project costs. In this case, the State Committee would pay down the project cost, which would reduce the loan from UAH100 to UAH50 (see Table 13.2).

As is clear, when comparing the "Total Paid by Local Government" in Tables 13.1 and 13.2, the *upfront grant results in substantially lower costs to the local government.*

Reducing the local governments' payments below UAH100 was not necessary. The mayors had agreed to pay an amount equal to the principal they borrowed. So, we suggested that the State Committee consider another strategy that would achieve the same result as in Table 13.1 (i.e., the local governments paying UAH100), but would substantially reduce the amount of the State Committee's grants.

In this case, the local government would simply pay an amount *equal to the payments on a 0% interest loan,* or 20 per year, as is the case in Table 13.1. The government would subsidize the project with an upfront grant, but only to the extent necessary to reduce annual principal and interest costs on the project loan to UAH20 per year—the amount agreed to by the local government (see Table 13.3).

TABLE 13.3
Reduced Upfront Grant

	Year						
	0	**1**	**2**	**3**	**4**	**5**	**Total**
Total Financing Required	100						
Subsidy by State Committee	40						
Outstanding Balance	60	52	42.4	30.88	17.06	0	n/a
Principal Paid by Local Gov't	0	8	9.6	11.52	13.82	17.06	60
Interest Paid by Local Gov't	0	12	10.4	8.48	6.18	3.41	40
Total Paid by Local Gov't	0	20	20	20	20	20	100

Now, in this case, please notice that the State Committee's grant has been substantially reduced, from UAH50 to UAH40. Thus, because of the subsidy, the outstanding loan balance is 60 on day 1.

In this case, assuming that the local governments were willing to pay an amount each year equal to the principal (but not interest) payment on a 5-year loan, the State Committee could accommodate the local governments while at the same time saving 10/50ths, or 20%, of its total project grant cost. These savings could then be used to subsidize additional projects.

So as you can see, when construction or acquisition of new facilities or other brick-and-mortar projects are involved, it is always better to make an upfront buy-down grant than to give annual grants to subsidize interest rates. *This is always true as long as the rate of inflation is below the rate of interest.*

When our project ended in Ukraine in 2007, the State Committee staff was still thinking about changing their program. I believe they still are.

Affordability

A year or so after the Clean Water Act was amended in 1987 to create the Clean Water State Revolving Fund (CWSRF), some of the thoughtful people at the US Environmental Protection Agency (EPA) began to wonder if communities could afford the new loan program. Grants were free but loans were not. Were communities across the country going to be able to repay these new loans?

So, they convened a meeting at the EPA for the specific purpose of discussing the "concept of affordability." Most people there were EPA folks, but they did invite a few of us outsiders. These discussions went very well, but there was one gentleman who sat there with his arms folded across his chest, squirming in frustration. Finally, no longer able to contain himself, he blurted out, "In the world of classical economics, there is no such thing as the 'concept of affordability,'" to which another of the participants gently responded, "Yes, but in the real world, there is no such thing as classical economics."

There are two affordability issues: *community affordability* and *individual affordability*. They are not mutually exclusive. You will have cases of individual unaffordability in the wealthiest communities. And, perhaps surprisingly, you will also have cases of individual unaffordability in places where there is community unaffordability as well, despite adjusting the rates.

We will deal with community affordability first.

Community Affordability

Back in the mid-1990s, the Environmental Finance Center at the University of Maryland convened a meeting to discuss the dilemma of a small town in western Maryland situated on a tributary of the Potomac River. Just a few feet below the surface of the land was a stratum of solid bedrock. This was a rural community with all of the homes on septic systems. But because of the bedrock below, the septic systems did not work very well. The scientists from the State were puzzled with the readings they were getting on their downstream water quality tests. After much testing and investigation, they concluded that the problem was coming from all of the town's septic systems on bedrock. So, the State ordered the town to install sewers. There were two problems with this requirement. First, to install the sewer mains, they would have to blast out the bedrock. And, second, the homes in this town were not very close together. They were going to need several hundred extra feet of sewer mains. So, this was going to be a very expensive enterprise.

This was not a wealthy town to begin with; back in those days, the average home value was under $100,000. The residents, then, were understandably shocked when the engineers concluded that the project would cost $39,000 per home!

This is a classic case of where a grant is needed. It is a classic case of community *un*affordability.* (And it is not classical economics!)

There is a rule of thumb in the environmental infrastructure business that says that rates of 1% of Median Household Income (MHI) for drinking water and 1% for wastewater are affordable. So, 2% for water and sewer combined. Notwithstanding the general consensus about the 1% each affordability number for water and sewer service, the EPA thinks it should be 2% each. Regardless of what the EPA thinks, most people think 1% for each is a good number; so that is what we will use.

As a matter of fact, the American Water Works Association (AWWA), in partnership with Raftelis Financial Consultants, Inc., undertook their "2010 Water and Wastewater Survey" of 49 states and the District of Columbia. With 308 responses for water and 288 for wastewater, the report found that the average water charge was 0.66% of MHI and wastewater charges averaged 0.84% of MHI. The total then is 1.5% for both. So, our 2% is a conservative number. It is a conservative number in one sense: that of measuring a community's ability to pay, not an individual's.

In Chapter 15, you will read about the "Curse of Subsidies." In the best of all possible worlds, the first thing we should do is to get rid of all the subsidies. The second thing we should do is to recalculate the cost of utility service against MHI. This is not going to happen. Most of the subsidies that environmental utility systems enjoy are associated with project financings.

* The State of Maryland did, in fact, address this problem with a substantial grant for the community.

As noted in Chapter 15, simple, old, tax-exempt municipal bonds are a form of subsidy. Furthermore, all of the state CWSRF programs offer generous subsidies. Most lend at approximately 50% of the going market rate. So, when AAA tax-exempt municipal bonds are selling at 5%, the most CWSRFs are making loans at about 2.5%. Does anybody think that any environmental utility is going to give up these subsidies? No, of course not. So, we need to settle for what might be called a modified full cost pricing.

The failure to set full cost rates is frequently attributed to community affordability limits, which may be wrongly characterized in terms of the community's ability to pay. In fact, the underlying cause is generally the existence of relatively large numbers of households with individual affordability problems. This creates a political climate in which significant increases in rate level are seen as unacceptable, due to the harm that would be visited on the lowest-income customers. (This situation can, of course, be effectively dealt with by targeting subsidies to those households with true need.)

However, in the absence of unlimited outside subsidies, utilities generally deal with this situation by using such ill-advised strategies as deferring maintenance, deferring facility upgrades and replacements, eliminating staff functions, and maintaining low wage levels. One of the most common manifestations of these funding problems in older American cities is a deteriorated water or sewer system that is plagued by leaks and overflows.

Individual Affordability

As Chapter 15 describes, the third thing we need to do, is identify all those near the bottom of the income curve who cannot afford a service and set up a subsidy for those poor households only. Even in Loudoun County, Virginia, with its $119,000 MHI, there are undoubtedly some families that cannot pay their water and sewer bills. In most cases, as you will see, when we stop the subsidies for the many who do not need them, we will have more than enough money to subsidize the few who truly do.

That said, having identified those who cannot afford their utility service, we now need to address this individual affordability issue by identifying appropriate subsidies to help them pay their bills. There are at least four choices here:

1. The subsidy can come from general tax revenues, through some established state or local program. As you will see in the example below, we have identified 4,000 households (out of 100,000) that cannot afford their utility rates. They each need a subsidy of about $20 per month, or $240 per year, for a total of $480,000. In this first case, the subsidy would come from the state or local government, or both,

out of their general revenues. In other words, the local or state government would take $4.8 million out of their general fund and pay it to the utility.

2. The subsidy could come from a specific program created by the federal government or a state government. The model for this method is the Low Income Household Energy Assistance Program (LIHEAP), which helps the poor pay their energy bills. The federal or state government could create a similar specific program for those with problems paying their water and sewer bills. In this program, the government provides the subsidy directly to the consumer, who in turn pays the utility. In many cases, those who would use such a program would be the same people as those who use the LIHEAP. The federal government could also just expand the LIHEAP from energy only, to any type of utility bill. LIHEAP certainly is not a euphonious acronym, but it is better than LIHUAP.
3. The subsidy can be a cross-subsidy. In this case, the ratepayers would be divided into two categories: hardship and non-hardship. In this case, the subsidy for the hardship ratepayers comes from the non-hardship ratepayers. This method is illustrated below.
4. In this case, the subsidy can come from all utility customers, whether or not they can be considered hardship customers.

Each of the above four options has advantages and disadvantages. The first two options (subsidy from tax revenues and federal programs) may not be possible or feasible. However, there is the LIHEAP as a model; so we need not completely rule out the possibility that the federal government or a state might make financial assistance available, either from general revenues or from a specific program.

The difference between the last two options can be illustrated by a very simple example. This example considers only the case where there is no outside subsidy available to the utility—the program is funded by increasing some charges, or charges to some, or all, customers. The example is constructed for a medium-sized community of 100,000 households. We will assume that the MHI for this community is about $50,000, which is about the national MHI. We will also assume that the MHI for our hardship cases is about $20,000, which is below the poverty level for a family of four.

Using our rule-of-thumb affordability number of 1% of MHI, the community in general should be paying about $500 a year. In our example, they will pay $40 per month, which is $480. Close enough.

Our hardship cases should be paying only $200 per year. But in our case, we are going to target a subsidy of $20 per month for them. This means, of course, that they should be paying $20 per month, or $240 per year. This is just above 1% of MHI, actually 1.2% of MHI; but so be it.

As you will see below, our system has 100,000 ratepayers, of whom 4,000 are hardship cases. So, to create a $20-per-month subsidy for the hardship cases, we need a subsidy of $40,000 per month, or $480,000 per year.

Assume: Number of (residential) customers = 100,000
Number of hardship customers = 4,000
Total water sales = 750 MG per month
Price of water = $2 per 1,000 gallons
Average water sales per household = 7,500 gallons

A. *Base case*: The original, subsidy-free rate design. Hardship customers pay the exact same bill as do all other ratepayers.
B. *Targeted subsidy, financed by non-hardship customers*: This design incorporates a subsidy of $20.00 per month for hardship customers only, financed by increasing the fixed charge for non-hardship customers. Note that the increased cost for non-hardship customers is small.
C. *Targeted subsidy, financed by all customers*: This is the same as design B, except that all customers contribute equally to financing the subsidy. The result (as compared to B) is a slightly lower cost for non-hardship customers and a slightly higher payment (as compared to B) for hardship customers.

	A	B	C
Non-hardship customers:			
Fixed charge ($/month)	25.00	26.22	25.40
Variable charge ($/1,000 gal)	2.00	2.00	2.00
Monthly variable charge ($)	15.00	15.00	15.00
Average bill ($/month)	40.00	41.22	40.40
Hardship customers:			
Fixed charge ($/month)	25.00	5.00	5.40
Variable charge ($/1,000 gal)	2.00	2.00	2.00
Monthly variable charge ($)	15.00	15.00	15.00
Average bill ($/month)	40.00	20.00	20.40

As you can see, the impact of both alternatives B and C for the non-hardship customers is totally minimal. Likewise, the subsidy is extremely valuable to the hardship customers. Even in alternative C, the hardship customers' monthly bills are only 2% higher than in case B.

Whether or not you believe in classical economics, affordability is a real issue that needs to be dealt with. It is one of the four legitimate uses of grants.

If the issue is that a given community simply cannot afford to pay the cost of the ongoing, day-to-day operations of its water or sewer system, then the county and the state, and maybe even the federal government, have a real problem on their hands.

If the cost of a project makes rates unaffordable, then the project cost can be bought down by a one-time, upfront grant.

If the problem is that certain users of the system cannot afford their monthly bills, then these people must be identified, and a subsidy (either internal or external) must be identified and used to assist those needy households.

14

The Role of Equity

Grants, or gifts, have no monetary cost to the recipient. They are, therefore, the least cost method of financing environmental improvements. *Equity*, on the other hand, is the single most expensive method of financing environmental improvements. This is somewhat ironic. In classical corporate finance, the rule may be just the opposite. If you are the CEO of a large publicly traded company, you might observe that the best loan (or corporate bond) rate you can get is 5%, which means that you can borrow $100 in return for an annual payment of $5, or 20 times your money's worth. At the same time, you might observe that your shares are selling at 30 times earnings. In this case, if you really need money, you would issue more shares, paying your (equity) shareholders $3.33 rather than paying bondholders or a bank $5.

What exactly is equity? *Merriam-Webster* gives three definitions,* the first of which is "justice according to natural law or right." The second definition deals with the doctrine of equity under the common law. The third definition concerns finance. It is: "the money value of a property, or of an interest in a property, in excess of the claims or liens against it."

What are the "claims or liens" that could lie against a property? Essentially, they are the claims of creditors, that is, those to whom the "property" owes money. It is important to remember that *all* creditors get paid, and paid *in full*, before any equity (or share) holder gets paid a penny.

In corporate finance, creditors are divided into two general classes: *secured* and *unsecured* creditors. This is not just a rhetorical distinction. It goes directly to the question of who gets paid first. This becomes especially important if there is not enough money to pay all the creditors what they are owed. If a "property" has a value of $100 but has $90 owed to secured creditors and $20 owed to unsecured creditors, then the secured creditors will get paid 100¢ on the dollar on their claims. They will get all of their $90 back. The unsecured creditors, on the other hand, will get stiffed. Instead of getting their $20 back, they will only get what is left: $10. They will only recover 50¢ on the dollar on their claims.

Unsecured creditors are generally vendors and tradesmen, but also include professionals such as lawyers, engineers, and technical consultants to whom money is owed for services rendered.†

* Cf. http://www.merriam-webster.com/dictionary/equity.

† Certain categories of otherwise unsecured tradesmen, such as building contractors, carpenters, electricians, et al., may be given preferential treatment under state statutes, where money owed them can be secured by devices called *mechanics liens*.

The category of *secured* gets further subdivided into "senior" and "junior" creditors. These terms have nothing to do with age. Rather, as you might imagine, they have more to do—again—with priority of payment, that is, who gets paid first. It is the "seniors" who get paid first. Then, after the "seniors," the "juniors" get paid. Then, after the "seniors" and the "juniors" get paid in full, the general (unsecured) creditors get paid. Then, after all the secured and unsecured creditors have been paid in full, and only then—if there is any money left—do the equity-holders get paid. So, you see, as far as priority of payment is concerned, equity-holders are at the very bottom of the food chain.

Remembering that, by now, the total dollar value of shares of common stock on all the stock exchanges in the world should be over $50 trillion, and that there are billions of people who each own tens of thousands of dollars of equity in real property, and bearing in mind that equity-holders get paid last, why would anyone in their right mind invest in equity?

The answer is one word: Rewards!

Let us say you put a $50,000 down payment (equity) on a $250,000 home and get a mortgage (secured loan) from a bank for $200,000. Then, perhaps a few years later, you sell the house for $350,000. Well, the bank gets their money back, that is, whatever is left unpaid on the mortgage plus interest up to the date of sale. You, on the other hand, get your $50,000 equity investment back but you also get another $100,000, which is a 200% return on your investment. If the bank was charging you 5%, you can certainly say that you made out significantly better than they did.

On the other hand, imagine that you bought that home in 2006, before the sub-prime mortgage crisis. In such case, after the crisis hit, when you went to sell it, you learned that you could only get $150,000 for it. In this case, the bank gets all $150,000 (which leaves them short). And you are wiped out; your entire $50,000 equity investment is gone.

So, what is equity all about? A high risk of getting wiped out (or losing serious money) and receiving great financial rewards for taking that risk.

Think about buying a $100 corporate bond and one share of common* stock of a company whose shares are also selling for $100. Let us say the bond pays 5%. At the end of a year, the bond is redeemed and you get $105. But let us say that during that year, the stock price goes from $100 to $150. In this case, you would have made $50 on your investment, which is 10 times the $5 return that you would have gotten on the bond. Your risk is that the company would not go bankrupt. If you are willing to take the equity risk, you get a big financial reward. Traditional finance is about risk and rewards. In general, high risk means high rewards and that means equity.

That is precisely why equity has such a limited role in environmental finance.

* Equity, at least the equity (stock) of publicly traded companies, is also subdivided into two categories based on risk: preferred and common. Yes, *preferred* shareholders get paid before *common* shareholders in the event of a bankruptcy or corporate liquidation.

You know that Environmental Finance Principle 2 is to drive down the cost of financing environmental improvements. A corollary to that principle is that we must drive down the risk of nonpayment inherent in financing environmental projects. The higher the risk, the more equity is needed; and, therefore, more money is needed to pay the higher returns demanded by equity investors. Paying high returns shoots the environment in the foot because it means that more money will be absorbed by one project, leaving less money to pay for other projects.

On the other hand, the lower the risk, the lower the payments. That is the goal of environmental finance: to drive down the risk so that project payments will be lower, thus leaving more financial resources available to pay for other projects.

In traditional corporate finance, the price of shares of corporate stock is obviously set by a very efficient market of millions of buyers and millions of sellers. In general, the price is based on the collective mass expectation of how much the company will earn in the future. The share price is a multiple of the current earnings. If the world expects that earnings will grow, the ratio of price to earnings (the P/E ratio) will be very high. If the company is expected to lose money, or have lower earnings, then the P/E ratio will be very low.

When some of the Internet giants first made the Initial Public Offering (IPO) of their shares, they had not yet made a penny of profit. Their initial share prices were based solely on the expectation that the company would make a profit and that the profit would be astronomical. At the other end of the risk/reward spectrum in finance is debt. Whereas equity is high risk, high reward; debt is generally low risk, low reward.

Interestingly, however, the *risk/reward spectrum* is more than just a figure of speech. It actually is sort of a continuum, with the riskiest forms of debt rubbing elbows with the most conservative forms of equity. What are the most conservative forms of equity? Well, the shares of electric utilities in regulated states are probably a good example. After all, they sell an essential commodity: electricity. And, in a heavily regulated environment, they are most unlikely to go bankrupt. The same is true of the shares of publicly traded water companies, like American Water, which is 125 years old and serves 15+ million people in thirty states. American Water is most unlikely to go bankrupt.*

On the other hand, what are the riskiest forms of debt? The answer is junk bonds. What is a junk bond? A *junk bond* is debt that carries one of the lowest credit ratings awarded by the three internationally recognized credit rating agencies: Standard & Poor's, Moody's Investor Services, and Fitch Ratings. All three rating agencies use credit classifications. The top 10, using Standard & Poor's nomenclature, are called *investment-grade ratings.* They are: AAA,

* On the public side, many commentators say that the debt of public water/wastewater utilities—because they provide essential services—is more secure than the general obligation debt of local governments.

AA+, AA, AA–, A+, A, A–, BBB+, BBB, and BBB–. Any ratings of BB+ or below are *non-investment-grade ratings*, which denote speculative investments. Junk bonds are speculative investments. When Donald Trump financed his gambling casino, the Taj Mahal, he issued $1 billion of junk bonds to help pay for it. If low-risk corporate debt, at any given time, will command an interest rate in the 5% range, then a comparable junk bond would need an interest rate of 10% or above. It is no coincidence that when junk bonds have yields in the 10% range, the publicly traded equity shares of regulated electric utilities will also be yielding in the same range. Low-risk equity often trades close to high-risk debt. The risk spectrum is, indeed, a continuum. So is the reward spectrum.

Returning to the concept of equity, in addition to the shares of publicly traded companies, there is another form of equity: private equity. *Private equity* is the type of equity found in environmental finance projects. Private equity is the domain of major financial institutions and of very wealthy individuals. It is not available to the general public. It is certainly not for the faint-hearted. Unlike the shares of publicly traded companies, private equity is very illiquid. This means that once you invest, it is very difficult to exit the investment until you are paid out or wiped out.

Most private equity is invested in the form of venture capital. These investments are typically in new companies, new products, or even whole new technologies. Generally speaking, the number of times the word "new" is used (implying risk) in describing the venture, the higher will be the reward expected. In the early 1980s in New York, I started a company to provide financial guaranty insurance to industrial development bonds. The product was new. No one had ever insured an industrial development bond before. There were several new concepts involved, such as an obscure legal measure called an "Event of Non-Compliance." And, of course, the company was new. The only thing that was not new was me. I had managed the industrial development bank for the State of New York. So, on Wall Street at least, I was a semi-known commodity. But everything else was new. New, new, new. And the newness was reflected in the high returns that most of the venture capitalists wanted for their investments. Most wanted returns in the 20% to 25% range.

Private equity is also the type of equity found in environmental finance*—certainly in environmental project finance. Some of the best examples of private equity, in the context of environmental finance, involve investments in renewable energy and urban mass transport.

First, let us look at the type of Return on Investment (ROI) that equity investors seek. As you saw, an investor in the shares of a publicly traded electric utility would probably want a return, under normal circumstances, in

* Of course, there are also shares of publicly traded companies, such as American Water, that perform environmental services.

the 10% range. This is at the low end of the ROI range for equity—consistent with the relatively low risk involved in owning such shares.

At the other end of the risk spectrum, I like to think, is a Siberian gold mine. Not a mine that has been operating successfully for many years. Not even a mine that is operating at all. Rather, the new construction and start-up operation of a mine in deepest Siberia, where the only evidence of gold is some mineralogists' reports. New gold deposit. Foreign country. Very remote location. Forbidding climate. New company. New management. Untried labor force. Unknown technical barriers, etc., etc., etc. Do you doubt that any investor would accept less than a 40% ROI? Why so much? One word: Risk!

Here are some representative ROIs for private equity investments over a range of 3 to 10 years on a notional investment of $100:

	Years			
ROI	3	5	7	10
10%	$133	$161	$195	$259
15%	$152	$201	$266	$405
25%	$195	$305	$477	$931
40%	$274	$538	$1,054	$2,893

There you have it. If you are willing to invest in a new Siberian gold strike, you are going to want about $2,893 back for every $100 that you are willing to invest over a 10-year period.

Getting back to private equity investments in environmental improvement projects, a word of explanation is due here. As you might imagine, investments in renewable energy projects and in urban mass transport have an operating time envelope of certainly more than 3 years and even more than 10 years. There is a subtle, but fundamental, difference between these types of private equity projects and the more traditional private equity investments in new ventures. In venture capital, the payout is one-time and cumulative. In environmental project finance, the payout is annual.

The one-time cumulative payout in venture capital finance is called the *exit*. The way they get paid, or at least the way they think they are going to get paid, is called their *exit strategy*. Venture capital investors want a known, and relatively short (<10 years), investment period. They want out within 10 years. They want either to have an IPO of the new firm's stock, or to sell the new company to an industry giant (like a new Internet company being sold to Google or Amazon), or to be able to sell to another venture capital company. This is not exactly liquidity, but it is a way out. The exit strategy is as important to venture capitalists, as is ROI. After all, what good is a huge ROI if you cannot get your "I" back?

As far as private equity investments in renewable energy, urban mass transport, and other types of environmental improvements are concerned, there is no short-term investment horizon and there is no well-defined exit strategy. In these types of projects, the returns are annual: Once the projects get going, the equity investors get an annual payout.

Using a $1,000 investment as a model this time, here is what this would look like:

ROI	10%	15%	25%	40%
Payout	$100	$150	$250	$400

Recall that we said at the beginning of this chapter that equity was the most expensive method of financing? Well, let us compare the equity payout with a typical debt payout using a $1,000 investment again:

ROI	5%	10%	15%	25%	40%
Equity	n/a	$100	$150	$250	$400
Debt	$50	n/a	n/a	n/a	n/a

As you can see, debt is much less expensive than equity. Note also that there is no 5% return on equity because no equity investor would ever accept a 5% return. Likewise, there are no returns for debt above 5% because there is no debt at these interest rates (for well-conceived and -executed projects).

Climate change projects, like urban mass transport and renewable energy, are not financed with 100% equity. I have never heard of one being financed with even 50% equity. But at percentages below 50%, there is equity in a sizeable number of deals.

Let us now look in depth at the impact of equity on the annual payments required of projects.

Below are a series of tables that depict the flow of payments at various proportions of debt and equity. In each, the project cost will be $1,000, and the debt will be in the form of a level payment loan at an interest rate of 5% and a term of 30 years. For the sake of comparison, if the project contained 100% debt and no equity whatsoever, the total project payment would be $65.05.

The table below illustrates the payment flow when a project is composed of $600, or 60%, debt and $400, or 40%, equity:

ROI	10%	15%	25%	40%
Equity ($400)	$40	$60	$100	$160
Debt ($600)	$39	$39	$39	$39
Total Payment	$79	$99	$139	$199
Payment on 100% Debt	$65	$65	$65	$65
Equity Premium	$14	$34	$74	$134
% Increase	21.5%	52.3%	113.8%	206.2%

As you can see, injecting 40% equity into a deal substantially increases the amount of money required each year to pay for the project.

Here is the same presentation, but this time with 30% equity, or $300, and 70% debt, or $700:

ROI	10%	15%	25%	40%
Equity ($300)	$30	$45	$75	$120
Debt ($700)	$46	$46	$46	$46
Total Payments	$76	$91	$121	$166
Payment on 100% Debt	$65	$65	$65	$65
Equity Premium	$11	$26	$56	$101
% Increase	16.9%	40.0%	86.2%	155.4%

Again, equity forces substantially higher payments. Now we present the same data with an 80%, or $800 debt, and 20%, or $200, equity:

ROI	10%	15%	25%	40%
Equity ($200)	$20	$30	$50	$80
Debt ($800)	$52	$52	$52	$52
Total Payments	$72	$82	$102	$132
Payment on 100% Debt	$65	$65	$65	$65
Equity Premium	$7	$17	$37	$67
% Increase	10.7%	26.2%	56.9%	103.1%

Now, here we are, finally, with a 90/10, debt/equity, ratio:

ROI	10%	15%	25%	40%
Equity ($100)	$10	$15	$25	$40
Debt ($900)	$59	$59	$59	$59
Total Payments	$69	$74	$84	$99
Payment on 100% Debt	$65	$65	$65	$65
Equity Premium	$4	$9	$19	$34
% Increase	6.2%	13.6%	29.2%	52.3%

As you can see, even with a relatively modest 10% equity share in the project, the electricity or the bus/train ticket will cost more because the annual project payments are higher.

In general, when I compare project costs, I generally use an 80/20 debt/equity split, with the equity requiring a 15% return. When making proposals to potential equity investors, these numbers seem reasonably acceptable. That said, you can see that putting 20% equity into the deal with a 15% ROI increases the annual project cost by over 26%. Because Principle 1 of environmental finance is to produce the greatest benefit for the largest number of people at the lowest possible cost, we must adopt the following rule: Because equity always increases project cost, we should never use equity funding in a project *unless it is absolutely necessary*. That brings us squarely to the following question: When is it absolutely necessary to inject equity into a project?

The answer has to do, odd as it might initially seem, with expertise. It has to do with Public–Private Partnerships (PPPs). It has to do with what the private equity pros call "having skin in the game."

There are many circumstances where there are private-sector companies that have great expertise and experience in building and operating projects such as wind power, solar power, Metro, Light Rail Transport (LRT), or Bus Rapid Transport (BRT) systems. When companies like these are organizing such projects, there are two distinct groups who want to see these companies invest—substantially—in their own projects. These two groups are the debtholders (i.e., lenders or bondholders) and local governments that own the projects, when it is a Metro, LRT, or BRT type of project within the local government's jurisdiction. This is the case of a PPP.

Take the City of Hanoi as a theoretical example. Hanoi might be in serious need of a mass transport project. They may also not feel that they have sufficient expertise with such projects to undertake them on their own. In such case, Hanoi might advertise or solicit bids from private-sector companies to enter into a PPP with the City of Hanoi to undertake a BRT project. In such case, the City would certainly check out the professional credentials of each potential partner. They would probably seek the assistance of, or at least confer with, other entities such as the Vietnamese Ministry of Transport, or the Asian Development Bank or the World Bank. But, even so, once the City was satisfied that their private partner was competent to build and/or operate their proposed BRT system, they would want to assure themselves that the project will ultimately be successful. They would want to make sure that their private partner does not just pay themselves huge fees, build a shoddy project, and then vanish. One of the most effective ways of ensuring this is to *require* the private partner to invest its own money in the project. Furthermore, such investments must be in the form of equity (or otherwise subordinated to all other investors). Hanoi will want to make sure that if their project is unsuccessful, that their private partner will lose a great deal of money. Equity is the riskiest form of investment; so Hanoi will want to make sure that their private partner takes the riskiest position.

The downside to this requirement is that the private partner will want an equity-style ROI. In these cases, the ROIs generally run in the 10% to 15% range. So, the bad comes with the good. The people of Hanoi will pay higher bus fares or the City will pay higher subsidies because they need to pay an equity premium to their private partner.

One way that equity payouts are facilitated is to organize the debt into *level principal payment loans/bonds*. This is a bit intricate. As you know, with *level principal payment bonds/loans*, the amount of interest to be paid each year, decreases as does the total amount of the annual debt service payment. And so, in this case, the equity partner gets the benefit of the ever-declining debt payments. Here is how this works. We will use a theoretical (and highly improbable) five-year finance term to make our point.

Let us use a $1,000 project with 20% or ($200) of equity from the private partner/builder-operator, and 80% (or $800), of debt for a term of 5 years at a rate of 5%, repaid according to the Level Principal Payment method. Looking at the table below, line 1 is the level principal of $160 per year for five years, totaling the $800 of debt invested. Line 2 is the declining annual interest payment. Line 3 shows the declining annual debt service payments. Line 4 is what is left, assuming that the system has Cash Available for Debt Service (CADS) of $200 per year. The money in line 4 goes to the private equity investors. Line 5 shows the discounted value of these payments, using a 3% discount rate (an assumed rate of inflation). Line 6 shows the cumulative ROI at the discounted rate, and Line 7 shows the private equity investors' ROI as a percentage.

	Year				
	1	2	3	4	5
Debt ($800)					
Principal	$160	$160	$160	$160	$160
Interest	$40	$32	$24	$16	$8
Total Debt	$200	$192	$184	$176	$168
Equity ($200)	$0	$8	$16	$24	$32
Discount @ 3%	0	$7.54	$14.64	$21.32	$27.60
Cumulative	0	$7.54	$22.18	$43.50	$71.10
% Cumulative ROI	0	3.8%	11.1%	21.8%	35.6%

A cumulative, fully discounted ROI of 35.6% is a truly fabulous return for the private partner/builder-operator. But just remember the statement that if the $1,000 project were financed with 100% debt, the annual payment would be $65.05. In our private partner/builder-operator scenario above, the annual project payment is $200. Ouch! The payment is more than 300% higher—just to make sure the builder-operator does a good job. What a price to pay!

Now you may notice my statement that the PPP example has a 5-year financing term, which is abnormally short. This is certainly true. A 30-year term on the debt is what you would expect. Or, 20 years at a minimum. Here is what the scenario would look like—for the first 5 years—if there were a 20-year debt.

	Year				
	1	2	3	4	5
Debt ($800)					
Principal	$40	$40	$40	$40	$40
Interest	$40	$38	$36	$34	$32
Total Debt	$80	$78	$76	$74	$72
Equity ($200)	$0	$2	$4	$6	$8
Discount @ 3%	0	$1.89	$3.66	$5.33	$6.90
Cumulative	0	$1.89	$5.55	$10.88	$17.78
% Cumulative ROI	0	0.95%	2.78%	5.44%	8.89%

Notice now that the private partner is earning significantly less—perhaps so little that he will decline to participate in the PPP.

Is it any wonder why—under this scenario—the private partner will always want the shortest debt term?

Even with this reduced scenario, please note that the first year's debt service is $80, which is 23% higher than the $65.05 that it would cost if the project were funded with 100% debt.

Please remember that using equity violates Principle 1 of environmental finance. It does *not* provide the greatest environmental benefit to the largest number of people at the lowest possible cost. Therefore, equity should be used only when it is absolutely necessary.

15

The Curse of Subsidies

The role of subsidies, as you may recall, was covered briefly in our earlier discussion of maximizing Cash Available for Debt Service (CADS). Here, however, we will focus on the unique and special role of subsidies in paying for environmental infrastructure.

If you looked at the full landscape of subsidies available for environmental infrastructure projects, you would be completely excused for being totally bewildered. When a normal person thinks of subsidies, he usually thinks of helping the poor pay for things they cannot afford. Right? Well, in the environmental infrastructure game, being poor has nothing to do with anything... most of the time. Take the case of Loudoun County, Virginia, which has the highest Median Household Income (MHI) in the United States. Their MHI was about $119,000 in 2011. That was the *median*. That means that 50% of the households earned *more than $119,000*. Let us now imagine the household with the absolute highest income in Loudoun County. Would it surprise you to learn that that family receives subsidies from the Commonwealth of Virginia for their wastewater services? Well, they do.

A very wise mentor of mine breaks down subsidies into two main categories. The first is *general, supply-based subsidies*. That is what you see in Loudoun County. The second category is *targeted, demand-based subsidies*. Remember in Chapter 1 my story about Jack and Diane? Do you remember that I said that the statewide $300 tax credit for installing a new, clean-burning wood stove was a waste of money? Instead, I said the State should have set up a 100% cash subsidy for people (1) living in non-attainment areas, (2) below the poverty line, (3) with no other source of home heat, and (4) with a resident with a pulmonary ailment. This is what I mean by a targeted, demand-based subsidy.

Now let me show you how truly wasteful general, supply-based subsidies are by illustrating the point with the most infamous example of all: the tax-exempt municipal bond. In 1895, the US Supreme Court ruled that Congress had no power under the US Constitution to tax the interest on state or local debt*. In 1913, this position became statutory.

When I was a young man in the municipal bond market, the rule of thumb was that the best (AAA) municipal bonds would have interest rates of about 75% of US Treasury bonds of the same term. So let us envision an interest rate

* This case, *Pollock v Farmers' Loan & Trust Co.*, was effectively overruled by the US Supreme Court in 1988 in *South Carolina v Baker*. But the statutory prohibition remains intact.

environment where, say, 20-year Treasury bonds carried an 8% interest rate. Under these circumstances, you would expect a AAA tax-exempt municipal bond to carry a rate of about 6%. Now let us say that the 8% Treasury bond was bought by the richest householder in Loudoun County. We will call her Ms. Loudoun. If she bought $10,000 of these bonds, she would receive $800 a year in interest. Being in the 35% tax bracket, Ms. Loudon would pay $280 in federal income taxes and thus have net earnings of $520.

Now let us say, instead, that Ms. Loudoun bought $10,000 of tax–exempt bonds at a 6% interest rate issued by the Commonwealth of Virginia. In this latter case, Ms. Loudoun would receive $600 per year in interest. Because this interest is exempt from federal income taxation, Ms. Loudon gets to keep the entire $600. She does not pay a penny of that $600 in taxes.

So, we now know that if Ms. Loudon bought the 6% tax-exempt bond, she would be $80 better off than if she had bought the 8% Treasury bond and paid taxes on it.

But what about the $280 that the US Treasury *did not get, that they missed out on*? Let us look at it this way. There are 535 members of Congress. Each one receives a salary of $174,000 per year. That means there should be $93,090,000 in the Congressional payroll account. Let us say that this account is $280 short because Ms. Loudon opted not to buy the Treasury bond. If this were really the case, panic would sweep Capitol Hill! Undoubtedly, Congress would immediately introduce a bill to raise some tax by $280 to make up the shortfall. Who would pay that tax? You and I would—as would the poorest tax-paying family in Buffalo County, South Dakota, the poorest county (by MHI)* in the country.

Is this your idea of a good subsidy—where you and I and the Buffalo County folks make up for the subsidies effectively paid to Ms. Loudoun? I say "effectively" because Ms. Loudoun did not actually save the $280; she saved $80. What happened to the other $200? Who got that savings? The Commonwealth of Virginia did. If there were no tax-exempt bonds, Virginia would have had to pay $800 instead of $600. So, the taxpayers of Virginia saved $200 and Ms. Loudon saved $80, and you (non-Virginians) and I and the people in Buffalo County made it up.

Subsidies are often set and left in place for years, regardless of the intervening growth in income of those using the service. Often, subsidies are left in place long after there are substantial rises in income for the general population. A prime example of this would be the high subsidization of the US farming industry. According to the Environmental Working Group, a major national environmental health research and advocacy NGO,

> From 1995 to 2012, Congress appropriated $292.5 billion in farm subsidies. The top 10% of farmers collected 75% of all subsidies, totaling $178.5 billion. These top 10% received average annual payments of $32,043. The

* The MHI in Buffalo County, South Dakota, is $12,692.

> bottom 80% of recipients collected an average of $604 per year, while 62% of US farmers did not receive any subsidies.
>
> Thus, while major and highly profitable commercial or industrial farms received massive subsidies, smaller and less-profitable farms (such as those not owned by major corporations) received relatively little subsidy. In other words, while the major farms received subsidies they (arguably) did not need, smaller farmers were left with either the crumbs or no subsidies at all to help them get through the year.

Now, the purpose of this chapter is twofold: (1) to expose the wastefulness of general, supply-based subsidies, and (2) to tell you how targeted, demand-based subsidies can be put in place. You and I and all of our like-minded colleagues are not going to get the municipal bond tax exemption repealed, nor are we going to bring an end to wasteful farm subsidies. However, my hope is that whenever you are elected to a board or next attend a public meeting where subsidies are discussed, you will strike a blow for ending wasteful general, supply-based subsidies and instead champion more cost-effective, targeted, demand-based subsidies.

But what does all this mean for renewable energy—and a sustainable environment specifically?

Achieving a quality and healthy environment is an expensive undertaking and, much like the issue faced by the US farming industry, environmental issues are frequently faced with hugely wasteful or misplaced subsidies.

Take, for example, the State of Maryland's recent attempt to enact a bill for stormwater treatment. In 2011, Maryland, with only approximately 5.5 million people, learned that their bill for stormwater treatment and remediation was going to cost them about $5.5 billion, or about $1,000 for every man, woman, and child in the state. Even if these funds were raised through a highly efficient, tax-exempt, municipal bond issuance, the cost would still be about $60 per person per year, or, with an average of 2.5 people per household, about $150 a year for every home in Maryland. (This is in addition to the $60 per year that each household is currently paying in the form of a Chesapeake Bay Restoration Fee.)

Moreover, because environmental quality—in this case, a "fishable, swimmable" Chesapeake Bay—is so expensive and so vital to the entire country, not a single penny should be wasted. Unfortunately, this is not the case. Subsidies are literally *everywhere*, tucked away into a wide array of programs. Even in the "highly efficient" municipal bond example above there is a hidden subsidy of about $7.25 per person per year, or about $18 per household per year. This hidden subsidy is meant to knock down the price of stormwater remediation from $168 to $150, resulting in an overall 12% discount or, more precisely, in a 12% subsidy. This is another example of a wasteful, general, supply-based subsidy.

What is the matter with this type of subsidy, you may ask? Well, this particular subsidy is embedded within the municipal bond issued to finance stormwater remediation projects, and the purchasers of these bonds do not pay federal (or state) income tax on the interest they earn. So, the US Treasury ends up being shorted funds. And who then must make up this difference? *We* do. But let me spell this out a little more clearly.

Let us say that these bonds were not just issued in Maryland. They were issued specifically in Howard County, Maryland. Howard County has the fifth-highest MHI in the United States; it is about $105,692 per household. But, despite their general ability to pay full price, because of this overarching 12% subsidy, instead of paying $168 per household, the people of Howard County only pay $150. They save the $18 per year despite the fact that more than half of them are earning over $105,000 per year. As was the case in Loudoun County above, because the income on the bonds is exempt from federal income taxation, this results in lost taxes to the US Treasury. Those funds must be made up. Someone must make up for the lost revenue. So, who pays? The people in Buffalo County.

Of course, it is not *just* the people in Buffalo County; it is all of us, all US residents. But you get the point: Why are the families in Buffalo County, who are only earning $13,000 on average per year, subsidizing the people in Howard County, who are earning roughly $100,000 per year? Subsidies like these are bad news; they are not only inefficient, but unfair as well.

So, the next question: Are there families in Howard County that cannot afford to pay $168 for stormwater? Absolutely. There have to be. Do they deserve a break? Yes, of course, they do. There is always a segment of any population whose income does not keep pace with the growing economy and thus will need subsidies in order to get by. And it is *this* population that should be subsidized, and this population alone, if we wish to allocate resources as efficiently and usefully as possible. But remember that the *median* is the mid-point. So, what about the people in Howard County who *are* way on the high end of the income spectrum? What about those who are earning not just $100,000, but several hundred thousand dollars a year? They certainly do not need anyone to help them pay for environmental health bills and programs with subsidies. Those who can afford the service should pay for it. However, we must also be cautious not to confuse these points by declaring that "All farmers should be without subsidies" or that "All farmers should receive subsidies." Neither is true. What is true is that "Some people need subsidies at least some of the time." *And this is precisely one of the problems with environmental finance: badly needed money is constantly wasted due to some unnecessary and unfair subsidies.*

These are the supply-based, general subsidies. These are the subsidies we must be concerned with.

Rather than be content with these general subsidies, we must work to instead tailor our subsidies so that they only go to those who truly need them and where they can do the most good. It is all a matter of designing a subsidy properly so that it targets only those truly in need. Going back to the example of Jack and Diane, it might actually cost the state less to create this type of a subsidy program. And, bearing in mind Core Environmental finance Principle 3 (Use carrots and sticks together), the state might accompany its subsidy program with a regulation *requiring* all households in non-attainment areas to get new wood stoves.

With these types of subsidies—demand-based, targeted subsidies—we can find the poor people in places like Howard County who cannot make their payments.

In the case of US farms, direct payments are tied directly to acreage: the largest farms get the largest subsidies. Now, there may be some very large farms owned by some very poor people, but there certainly are not very many. And even for those few, a targeted subsidy, as opposed to a general subsidy, will work.

The starting point for any financial discussion of environmental subsidies is twofold. First, a utility must absolutely minimize its cash expenses so that it only needs to collect a minimum amount from its users. Second, once the minimum cash requirements are known, a tariff must be designed to bring in the necessary income to pay all expenses plus any additional funds needed for debt service. This is called a *full cost recovery tariff.* Once such a tariff has been designed, the question of how to deal with the inability of certain truly needy users to pay their share of the tariff can be addressed.

Throughout the remainder of this chapter, we will discuss the following four keynote issues that are the major factors in formulating policy for subsidies for water and wastewater services:

1. The need for subsidies
2. Delivering the subsidy
3. Targeting mechanisms
4. Affordability measures

The Need for Subsidies

In every water system, for example, there will be some families that cannot afford to pay the market price for water services. However, because water is a necessity of life, it is the responsibility of the community or government to make certain that such families are given adequate access to water services despite their inability to pay. There are many affordability mechanisms available to decision makers who can choose the mechanism that is most appropriate for their water utility system.

The primary consideration is how much can the local population afford to pay? In the United States, this is measured using the MHI. In 2010, the American Water Works Association (AWWA) and Raftelis Financial

Consultants, Inc., concluded their "Water and Wastewater Survey" of 49 states and the District of Columbia. With 308 responses for water and 288 for wastewater, the report found that the average water charge was 0.66% of MHI and wastewater charges averaged 0.84% of MHI. The total, then, is 1.5% for both. So, for our purposes, 2% is probably a good number. In other words, if water and wastewater charges are more than 2% of MHI for a given family, they may need some assistance in paying for their environmental utility bills. The US Environmental Protection Agency (EPA) has set this standard at 2.5% of the median household income*. We will stick with 2%.

Once a standard of affordability has been decided, there are four other factors that must be considered in implementing an effective targeted, demand-based subsidy.

1. *Identifying the target population(s) for the subsidy:* It is one thing to say that anyone with a water/sewer bill of more than 2% of their MHI gets a subsidy. It is another thing entirely to figure out who is eligible. Families do not get a note from the government each year informing them of their MHI. So, how do you know what it is? What surrogates can be used for estimating MHI and, hence, an appropriate subsidy? Some surrogates are eligibility for certain programs such as food stamps, or being on welfare, or being eligible for the Low Income Household Energy Assistance Program (LIHEAP). Finally, there are the tax returns of the residents of the household. There are also some non-income-related targeting mechanisms that could be used. These are discussed below.
2. *Gauging the subsidy:* How can we implement a subsidy? Should any homeowner with an income below the household median get a 100% subsidy? How about if they are above 90% of the MHI? Should they get a 10% subsidy? If their income is above 80% of MHI, should they get a 20% subsidy. And, of course, if they were in the lowest 10% of the MHI, they would get a 100% subsidy. That is one way of doing it. There are many others. The utility board must decide how they will spread out the subsidy among their customers with varying (low) income levels.
3. *Keeping administrative costs low*: How do we cost effectively identify the actual people with the appropriate income levels? Having them prove that they are on welfare, or on food stamps, or eligible for LIHEAP should not be too burdensome. It may even be able to be done online. Tax returns can also be scrutinized. But then the utility

* The 2.5% standard for water rate affordability has been criticized for not being an accurate measure of affordability because it only captures the affordability of water for the middle-income groups; the affordability of rates for low-income groups (for whom affordability programs are targeted) is not captured in this measure.

has to pay someone to do this. Furthermore, there may be problems when there are several wage earners in the same household. Whatever the method selected for identifying the targeted subsidy recipients, utilities must be mindful of the cost of both this aspect of the program as well as the cost of delivering the subsidy to the eligible customers.

4. *Avoiding perverse incentives*: Subsidy design must consider the potential effects that the subsidy may have on the behavior of the recipients; for example, no subsidy should incentivize recipients to use water wastefully or to sell it*. There are many stories of those lucky enough to have a direct connection to a municipal water system selling water to their neighbors without such a connection. Or those where the water is priced very low, who will sell it to their neighbors who do not receive subsidies and, hence, pay much more.

Delivering the Subsidy

There are three primary ways of delivering a low-income subsidy to those eligible: supply side, demand side, and cross-subsidy.

1. *Supply side*: The state or local government or other external entity makes financial resources available to the utility to cover the deficit between the costs of service provision and the level of the water bill. In this case, resources are transferred directly to the utility and delivered to eligible customers through the tariff structure.
2. *Demand side*: Resources are given directly to the customers who are deemed to be eligible; this is usually done outside the tariff structure. In other words, all customers are charged the same amount of money for the same level of service, but the eligible customers are given a credit or rebate.
3. *Cross-subsidies:* Some groups of customers are charged more than the true cost of service provision, and this surplus is used to cover the deficit on another set of customers who then pay less than the true cost of service provision.

* There are stories of intrepid entrepreneurs in Haiti who somehow get access to a tank truck, go into the rich neighborhoods in the middle of the night, open the fire hydrants and fill up their trucks (there are no fire hydrants in poor areas), chlorinate it, drive it back to the poor areas, and sell it.

Supply-side subsidies are the traditional approach used to subsidize water utilities. However, experience now shows that they are not as effective as demand-based subsidies. With supply-side subsidies, the presence of major state transfers makes utility managers less concerned about controlling the cost and hence generates inefficiency. Supply-side subsidies tend to lower the general tariff level for all customers and thus often fail to reach the poor as anticipated. Hence, whenever subsidies are deemed necessary, the growing preference is for demand-side subsidies as they go directly toward covering the water bill of the poor household rather than covering the general budget support for the utility.

In practice, cross-subsidies and direct subsidies are not mutually exclusive, and a large number of public utilities use both simultaneously.

Non-Income Targeting Mechanisms

Other than by directly measuring income levels, there are three primary ways of identifying beneficiaries in order to target subsidies at those most in need of them:

1. *Amount of water used by a household*: Used with the increasing block tariff or with a lifeline rate (defined in the following section: "Affordability Measures") so that a small amount of water is provided below cost to all households.
2. *Characteristics of the household*: Indicators such as geographical location, type of dwelling, income level, or household eligibility for other government assistance programs.
3. *Self-selection*: The utility provides two distinct levels of service, a high-quality service at full cost and a low-quality service at subsidized cost with the assumption that only a genuinely poor person would choose the low-quality subsidized service.

The amount of water used by a household, however, is a weak indicator of poverty because poor households may actually consume relatively large amounts of water as many have comparatively larger families wherein all must share one pipe and there may be leaky appliances in need of repair. Geographical criteria only work in cities that have well-defined, localized areas of poverty. Of course, it is often the case that a large proportion of the poor do not necessarily live in slums but are scattered and thus render geographical criteria ineffective. Furthermore, the value of subsidies can be capitalized into property values and rents, and may thus be captured by landlords and not be extended to the poorer tenants as intended.

Affordability Measures

Given the important considerations above, local government officials and utility managers must choose the type of affordability measure(s) that will accurately target needy households for a subsidy that will give these households access to clean water while incentivizing them to more efficiently utilize the water from the utility system. There are four methods available for delivering a subsidy to poorer water customers: (1) lifeline rates, (2) refunds or discounts, (3) community social service funds, and (4) national funds for low-income water bill assistance.

Lifeline Rates for Low-Income Customers

Lifeline rates, in general, refer to a small amount of water (i.e., an amount for minimum essential use)* provided at a rate below the cost of service, while pricing higher-consumption blocks at higher rates. There are two basic types of lifeline rates: those that are applied to all customers for that small block of water used, and those that are only charged to customers who have proven their low-income status. Implementing a lifeline rate requires alterations in the pricing structure in order to separate blocks of water use for varying prices, in which case an increasing block tariff (IBT) rate must be created. Or, separate classes of residential customers must be identified: those that are eligible for the lifeline versus those that are not.

If the lifeline rate is limited to low-income customers, then it will achieve target efficiency (i.e., help only those who need it); however, if the lifeline is available to *all* customers, then the subsidy will be available to low-use customers who are not necessarily low income. If the latter were the case, then the lifeline rate would not be utilized effectively or fairly, as this policy would allocate the benefit to the wrong group of beneficiaries; in other words, to those beneficiaries who could have paid the full cost of the service themselves.

If lifeline rates are set below the actual cost of delivering the water, then there will be a cross-subsidy from those using more water. Of course, in this case, there would have to be an IBT. Otherwise, the utility would go bankrupt. For example, let us say that the cost of delivering 1,000 gallons of water a month is $1.50. Let us say that the lifeline rate is $1 for 1,000 gallons of water per month. But let us say that the next 1,000 gallons of water (i.e., from 1,000 to 2,000 per month) is $2 per month. That means that every customer who uses 2,000 gallons can effectively cross-subsidize one lifeline customer.

There is another good side to these IBTs: promoting conservation. To be sure, any volumetric charge for water encourages conservation to some degree. For example, a utility has a flat charge of $1 per 100 gallons, then if I

* "Essential use" varies widely from country to country. For example, in the United States, essential use is about 100 *gallons* a day per person. In Europe, essential use is about 100 *liters* a day.

use 200 gallons, I pay more. This encourages me not to waste water. However, if I paid $1 for the first 100 gallons and $2 for the next 100 gallons, my incentive to conserve would be that much greater. This notwithstanding, lifeline rates are sometimes criticized when they are targeted to low-income customers at below cost because other customers are forced to pay for the subsidy in the form of increased costs for their water consumption. This criticism can usually be overcome with transparency and by emphasizing conservation. When ratepayers fully understand that they must pay a little bit more to help the few in their community who really cannot afford to pay, and when they realize that the increasing block tariff also promotes conservation, their opposition should abate.

Refunds or Discounts for Low-Income Customers

As an alternative to providing a lifeline rate, many utilities prefer to direct subsidies to eligible low-income households by way of a refund or a discount on the specific households' water bills. The amount of the discount offered depends on the full cost recovery tariff required for the utility's revenue so that it may cover its costs and the generally accepted poverty level within the community. For instance, if the full cost recovery tariff is 2% of the MHI, a low-income person may be paying 5% or more of their income toward the water bill. Assuming that the local government and community members have decided on a maximum acceptable limit that a household should have to pay for water with respect to its percent of income, then the amount of the discount should equal the percentage difference between the maximum limit and the actual percent of household income that a low-income household would pay toward their water bill without the subsidy. For example, in the case where the full cost recovery tariff is equal to 2% of household income of, say $50,000, or $1,000 per year; this same $1,000 would equal 5% of the household income of someone earning $20,000 per year. In such a case, the system might charge $600 (or give a $600 rebate to) the $20,000 earner so that they also are paying 2% of their household income. The benefit to the utility of having discounts or lower rates for low-income customers is the increased likelihood of collecting payment from these customers; the subsidy makes it possible for these customers to pay more of their bills more regularly and promptly.

The important question that must be addressed once the amount of discount is determined, however, is who is going to pay for this discount? In order for the utility to continue to cover their costs of water delivery, they will have to charge all other customers higher rates in order to make up the lost revenue from the low-income discount (subsidy). Some local governments are willing to pay for such subsidies from their tax collections as a form of social service for needy households. These are supply-based subsidies.

Community Social Service Funds

The two previous methods required that a low-income customer subsidy program be established within the utility. However, the same goal can be accomplished without establishing such a program. A community social service fund can provide financial assistance to low-income households outside the realm of the utility. In this case, the utility can charge its low-income customers the same amount as it charges its other customers, but the low-income customers are then able to seek bill-paying assistance from the local social service fund. The utility's role in this program is to refer low-income customers to the community service fund in order to receive their assistance or subsidy. This usually occurs when a customer contacts the utility due to their inability to pay the water bill. Such funds can be supplied through social organizations such as Catholic Charities or the United Way.

National Funds for Low-Income Water Bill Assistance

Another possible alternative to these methods is the national program to provide bill-paying assistance to low-income families; such a program often works in a similar fashion to the local community social service funds. However, whereas a local community social service fund depends on other community members to serve the needs of all households within a single community, a national-level fund can help ensure that low-income customers in *all* communities have access to bill-paying assistance. Such a program would be particularly helpful in rural communities where many of the residents are eligible for low-income status and where there are comparatively few social service agencies. The Low Income Household Energy Assistance Program (LIHEAP) is just such a program. Through the US Department of Health and Human Services, the federal government delivered $3.47 billion of LIHEAP funds to the states in 2012. No such program exists at this time for water or wastewater; but it certainly could in the future, especially as wastewater/stormwater costs continue to increase dramatically.

Designing Better Subsidies

When designing subsidies, there are several criteria that must be taken into account. In the case of water tariff structures, for example, the four primary criteria are (1) genuine need, (2) accurate targeting, (3) low administrative costs, and (4) exclusion and avoidance of all perverse incentives. But how does one go about investigating and meeting these criteria?

When assessing genuine need, you must remember two key things: (1) Always question from the outset whether or not any particular group of water consumers actually merits a subsidy (and, if so, why); and (2) always study what percentage of household income is being spent on water and what people are able or willing to pay in order to enjoy improved water services. The US Department of Agriculture (USDA), in successfully managing its rural water and sanitation program since 1941, has also made it a policy to survey the water rates of surrounding communities when investigating affordability. Refer back, if you will, to our previous chapter on the value and importance of grants. When the USDA finds that a project is either not affordable *per se*, or that the project's annual debt service payments would put the utility's rates far in excess of those in neighboring districts, they then use a grant to buy down the cost of the project to affordable levels.

The USDA grant (subsidy) policy is not perfect, but it is far superior to the CWSRF (Clean Water State Revolving Fund) states that indiscriminately give subsidies to rich and poor alike.* At least the USDA investigates the income levels and rates of its applicants and their neighboring districts, and then targets its grants to those systems (not individuals) that actually need financial assistance in building their projects.

As for truly accurate targeting, one must go further. As we said before, even in $100,000+ MHI Howard County, Maryland, there will be some people who cannot afford an increase in their utility rates. The USDA does a good job of identifying financial need at the system level—but we must go below this level.

How, then, does one go about identifying the specific individuals who actually belong to this genuine need group that would become your target group? We have already discussed the most obvious of these. The first and most obvious is the ratepayer's tax return. A surrogate for tax data is if the ratepayer is already eligible for food stamps. A third means is LIHEAP. A fourth is welfare. In addition, each community has charity groups that work in the areas of low-income housing or low-income medical assistance. These groups will have their own means tests for determining eligibility for assistance. These can be evaluated and, if found acceptable, used. Finally, the wisest policy is probably to use more than just one test. Thus, if a ratepayer meets at least one of several of these income criteria, then that person could be eligible for water system subsidies.

The value of low administrative costs must also not be underestimated. While it is important to screen customers carefully for subsidy eligibility, the screening process can grow impressively expensive in administrative terms. Achieving balance, therefore, between the need for targeting accuracy and the associated administrative costs should be a primary goal of anyone seeking to effectively and efficiently design a subsidy. This goes hand-in-hand with the need to avoid all perverse incentives. For example, using water tariffs as a means of redistributing income between different customer groups can lead to serious conflicts with the efficiency objective, as it often introduces perverse incentives for households and industrial users to use or not use water.

* Some state CWSRF programs offer subsidies to all borrowers, but even deeper subsidies to communities with especially low MHIs.

Subsidies: Some Real, Some Imagined

As you might imagine, there are many kinds of subsidies. To illustrate this, we can look at the wastewater situation in Maryland. Consider, first, that for wastewater facilities, there are three different levels of treatment that carry three different subsidies.

The first—let us call it *basic treatment*—includes treatment plants as well as all the pump stations, sewer mains, and so forth that are needed to get wastewater to the treatment plant and then to the receiving body of water. In Maryland, basic treatment systems can borrow from the State Revolving Fund (SRF). All systems can borrow for 20 years at a subsidized interest rate of 50% of the going market rate. In other words, at this time, the State of Maryland can borrow for 20 years for about 4%. So, the SRF offers its borrowers a rate of 50% of that, or 2% interest. The annual payment on a 4% loan is about $74 for every $1,000 borrowed. The rate to the SRF's borrowers (2% for 20 years) results in an annual payment of about $61 per $1,000. So, the subsidy is $13 on a $72 payment, or 18% (13/72). If you are a "disadvantaged community," your subsidy is down to 25% of market rates, or 1% in today's world. Eligibility for such a subsidy is governed by Section 212 of the Clean Water Act (CWA) that limits funding to Publicly Owned Treatment Works (POTWs). In other words, no privately owned treatment works are eligible for this subsidy.

The second level of treatment, *Biological Nutrient Removal* (BNR), takes nitrogen down from 18 to 8 parts per million (ppm). Maryland provides a 50% subsidy for BNR in the form of a grant.

So, if you have a $10-million BNR project, you can get 50% (or $5 million) from the State—as a grant. Then you can borrow the other $5 million from the SRF at their subsidized rates. The State gets the $5 million for these BNR grants from its general revenues. Thus, it is a classic general, supply-based subsidy.

The third and final level of treatment is called *Enhanced Nutrient Removal* (ENR). This form of treatment takes nitrogen down to the recently mandated level of 3 ppm. Essentially, this mandate means that any wastewater facility that does not provide for ENR cannot get a National Pollutant Discharge Elimination System (NPDES) permit required to actually operate such a facility. For this level of treatment, the state of Maryland offers 100% grants and they are earmarked—at this time—for the 66 largest POTWs that discharge 95% of the point-source nitrogen. However, these grants will undoubtedly be expanded eventually to include all systems.

Now, the interesting aspect of these 100% ENR grants is that they are *not* funded from the State's general revenues; rather, they are funded through the "Bay Restoration Fee," which is actually a tax on every building with a toilet or drain in Maryland. Residences pay $5 per month, or $60 per year. Businesses pay more based on a formula. So you see, they are not funded from "general" tax revenues; rather, they are funded from a special tax. Furthermore, the Bay Restoration Fee, or flush tax as the wags call it, is based on either a building's connection to a sewer main or to a septic system. So the entire special tax is a wastewater tax. Instead of just telling the treatment plants to upgrade to ENR and then let them individually raise their rates by $5 a month, the State simply went ahead and did it on their own and then passed these "surrogate sewer rate increases" back to the system in the form of 100% grants. A unique procedure, as far as I know.

But before we move on, let me illustrate just how these various subsidies impact sewer rates. Let us take a $10-million project, for example: $5 million of this project budget is relegated to basic treatment, $2 million is dedicated to BNR, and $3 million is left for ENR (although it should be noted that ENR is typically the most expensive part of such projects rather than basic treatment). We are going to compare the subsidized costs that publicly owned facilities can avail themselves of, as opposed to the totally unsubsidized costs that a privately owned facility would have to bear.

	Cost (Private)	Subsidy	Net Cost to Ratepayers (Public)
Basic Treatment	$5,000,000	[b]	$5,000,000
BNR	$2,000,000	$1,000,000	$1,000,000
ENR	$3,000,000	$3,000,000	$0.00
Totals	$10,000,000	$4,775,000	$6,000,000
Year 1 Annual Cost to Ratepayers	$1,200,000[a]		$366,940[c]

[a] For *privately owned* facilities and based on a 12% Return on Investment (ROI), as permitted by the Maryland Public Service Commission, on a 30-year straight-line depreciation basis.

[b] The subsidy is reflected in a lower Net Cost to Ratepayers. The market cost of a $6,000,000 loan would be $441,490.

[c] For *publicly owned* facilities and based on a 20-year SRF loan at an interest rate of 2%.

In other words, the differences are vast. Ratepayer costs are about 325% *higher* with private ownership than with public ownership of water/wastewater facilities. Thus, while private firms may be better equipped and positioned to design, build, and operate water/wastewater systems, they are not necessarily the ones that ought to own the facilities. In states such as Maryland, with elaborate systems of grants and subsidies, it simply does not make financial sense for private firms to own these facilities, regardless of their abilities to design and operate them.

Reducing Costs

Now let us conclude this study of subsidies with a case study on reducing the costs related to them. For Maryland, there is a relatively easy way to reduce the cost of water/wastewater subsidies by more than 87%. The Maryland Water Quality Finance Agency (MWQFA) makes only 20-year loans because Section 603 (d)(1) of the CWA limits SRF loans to 20-year terms. Section 603(d)

(3) of the CWA, however, governs *financial guaranties* and there is no 20-year limitation in this subsection.

Thus, to finance water/wastewater projects in the state, Maryland sells bonds at an approximate rate of 4%. Thus, with a 20-year term for each $1 million raised, the annual debt service would amount to $73,582. A subsidized loan to a disadvantaged community at a notional rate of 1% with a term of 20 years would require a payment of $55,415. Thus, for every $1 million of funds the MWQFA lends to disadvantaged communities, it costs the state $18,167 ($73,582 – $55,415 = $18,167) in *annual* subsidies. Discounting this amount of money for 20 years at a presumed inflation rate of 3% would bring the total discounted value of Maryland's subsidy to $270,274 per $1,000,000, or more than 27% of total project cost.

Instead, Maryland could have the MWQFA *guaranty* bonds. The basis of the MWQFA's guaranty would be limited to its SRF assets—it would not involve the guaranty of the State of Maryland. A statewide agency could serve as a conduit issuer of tax-exempt bonds with 30-year terms for the MWQFA, and the MWQFA could then guaranty them. The MWQFA would still be in charge of the entire program; the state agency would simply issue bonds for projects approved by the MWQFA.

Now, the annual debt service on 30-year bonds, at the same 4% rate, would be $57,830. To achieve the same subsidy level as for the 20-year bonds, it would cost the State only $2,415 ($57,830 – $55,415 = $2,415) per million per year—instead of $18,167 per million per year. This is a savings of 87%, or $15,752 per million per year! The total cost of this subsidy to the State, discounted again at a 3% rate for 30 years, would be $47,337, or 4.7% of project cost—instead of $270,274—a difference of $222,937 for every $1 million loaned for water/wastewater facilities. Although the state agency bond would be 10 years longer, the critical number for the local utility borrower in these transactions is the annual payment—not the total payment—because it is the annual payment that directly affects their local water and sewer rates.

The table below illustrates these savings to the State of Maryland.

Bonds Issued By	**MWQFA**	**Local Wastewater Agency**
Amount of Bonds	$1,000,000	$1,000,000
Interest Rate	4%	4%
Term	20 years	30 years
Annual Debt Service	$73,582	$57,830
Subsidized Loan Payment	$55,415	$55,415
Maryland's Annual Subsidy	$18,167	$2,415
Maryland's Total Subsidy (discounted at 3%)	$270,274	$47,337

In 2011, there were 138 projects currently listed on the State's Project Priority List for the Clean Water Revolving Fund totaling $321,431,003, and 50 projects listed for the Drinking Water Revolving Fund totaling $135,768,884. Thus,

Maryland had some 188 pending projects totaling $457,199,887. To subsidize all these projects under the current program would have cost the State some $8,305,934 per year in subsidies, for a total of $123,569,003. If Maryland had changed policies to have the MWQFA guaranty 30-year bonds issued by a statewide agency such as the Chesapeake Bay Trust, it would only have cost Maryland $1,104,135 per year, or a total of $21,641,533. In short, if Maryland had changed policies, it would have saved $7,201,799 per year, or a total of $101,926,574. Thus, as important as it is to diminish administrative costs in designing a subsidy as much as possible, so it is vital that the methods used to fund such subsidies be as cost effective as possible. After all, the more cost-saving the methods utilized in the funding of one subsidy, the more likely it is that funds will remain to support another worthy project.

16

Leverage: The Power of Guaranties

There are really only three ways to convey money: (1) by gift (grant), (2) by loan (debt), or (3) by investment (equity). In Chapter 14, we discussed the uses of debt and equity. As you may recall, equity has a very limited role in environmental finance because of its high cost. Grant and loan programs dominate the environmental finance landscape.

Grants are pretty straightforward, as you will see in the next few pages. However, there are several important variants on loans, some of which are critically important. We are going to divide loans into three categories: (1) subsidized loans, (2) market-rate loans, and, perhaps surprisingly, (3) loan guaranties.

Subsidized loans are loans that are made at interest rates below market rates. In our comparison below, we use a 0% interest rate as an example.

Market-rate loans are just exactly what they mean. They are the rates that a qualified borrower could obtain by shopping his loan before several competent lenders.

Loan guaranties, or financial guaranties, are the oddball here. They are classified as "loans" because they involve traditional market-rate loans, but with one important additional feature: Their payments are guarantied by a financially powerful third party.

In this chapter, we will explore the power of financial guaranties. First, we will describe the four major means of financing projects: (1) grants, (2) subsidized loans, (3) market-rate loans, and (4) guaranties. Then we will set up a simulated test in order to compare them. But first a few words about these and some related concepts.

Why distinguish between market-rate loans and subsidized loans? Recall that the overriding purpose of environmental finance is to provide the greatest amount of environmental benefits to the largest number of people at the lowest possible cost. Then look at the comparison—later in this chapter—between market-rate and subsidized loans. In terms of our goal, subsidized loans are a terrible waste of resources. Subsidies, which are a form of grants, have their place but it is very limited. Chapter 15 is titled "The Curse of Subsidies." Subsidized loan programs are very popular for several very wrong reasons, so that chapter heading pretty much speaks for itself.

Noticeably missing from these financing mechanisms are municipal bonds. After all, the $3.7 trillion municipal bond market is, by far, the largest single source of financing for state and local governments. Indeed, municipal bonds are important, but they are actually a special form of market-rate loan so they will be discussed under that heading.

It hardly needs pointing out that these three means are not mutually exclusive. Obviously, a subsidized loan is both a loan and a grant.

So, please keep in mind that there are really only three ways to finance projects—grants, loans, and equity—and that subsidized loans, loan guaranties, and bonds, are just variants of "loans." This is important because one of the most persistent buzzwords in government is *innovative finance*. This is even true internationally. Every so often, you will actually see World Bank or United Nations documents refer to "innovative finance." This term is used as if it were some mysterious "other way to finance," which is misleading. All finance is grants, loans, or equity. When officials use the term "innovative finance," it usually means that they are trying to find a donor to give them a grant.

This chapter will have three basic sections. The first will present the concepts of grants, subsidized loans, market-rate loans, and loan guaranties. Second, it will present the pros and cons of each of these methods of financing. Finally, it will compare these four finance techniques from the perspective of how many projects—that is, how much environmental good—can be done by each method with identical sums of money.

Major Techniques for Financing Projects

There are four major techniques for financing environmental projects:

1. *Grants*: Sums of money awarded to finance a particular activity or project, which do not need to be repaid. This includes "soft" loans, which are loans that turn into grants if things go wrong.
2. *Subsidized loans*: A loan made to a qualified borrower at below the current market rate of interest.
3. *Market-rate loans*: A loan made to a qualified borrower at the current market interest rate.
4. *Loan guaranties*: A promise from a guarantor to make payment to the lenders in the case of nonpayment by the borrowers.

In creating or establishing a financial system for the funding of environmental projects, governments must choose which of the above, or which combination of the above, funding mechanisms it will use to finance projects. Each financing option has positive and negative aspects that must be considered when deciding which to incorporate into an environmental finance program. The most important factor for government officials to consider when making project finance decisions is the number of projects that can be funded with the limited resources available; this will be referred to as *efficiency*.

The concept of efficiency, as it relates to project finance, refers to the number of projects that can be funded with fixed or limited financial resources. High efficiency occurs when the most projects are funded with the least amount of money.

The following simulation will help demonstrate the impact on efficiency under different finance scenarios:

> Assume the existence of a "National Environmental Fund" (NEF), which is a government agency that has seed money in the amount of $100 million. The effect of the NEF project finance decisions will be simulated over a 10-year period, starting in Year 0. Each project is assumed to cost $5 million. The other rules of this game will be presented after the four financing methods are discussed.

Grants

The essence of a grant is that it does not have to be repaid. It is a gift. Because grants are not repaid, they improve cash flow for the recipient. In doing so, money that otherwise would have been spent on loan repayments can be put to other uses for the project. Grants are made to facilitate projects that otherwise are not affordable. These benefits come at some cost to the project managers because grantors often attach specific project requirements and conditions along with the funding, which limit project managers' use of the grant money. Projects are often required to meet the specific goals of the granting governmental agency or private organization. From the perspective of the funding agency, a grant is the least efficient use of fixed capital assets because once the grant is made, the money is gone and is not repaid.

> In our first simulation, the NEF managers choose to fund all environmental projects with grants. So, they make twenty grants in the first year. That is all. There are no funds for grants in subsequent years. Table 16.1 shows how it works.

TABLE 16.1

NEF Subsidized Loan Program

	Year										
	0	1	2	3	4	5	6	7	8	9	10
Opening Balance	100	0	0	0	0	0	0	0	0	0	0
Interest Earned	0	0	0	0	0	0	0	0	0	0	0
Principal Repaid	0	0	0	0	0	0	0	0	0	0	0
Total Funds Available	100	0	0	0	0	0	0	0	0	0	0
Amount of Grants Made	100	0	0	0	0	0	0	0	0	0	0
Ending Balance	0	0	0	0	0	0	0	0	0	0	0
New Grants Made	20	0	0	0	0	0	0	0	0	0	0
Total Grants Made	20	0	0	0	0	0	0	0	0	0	0

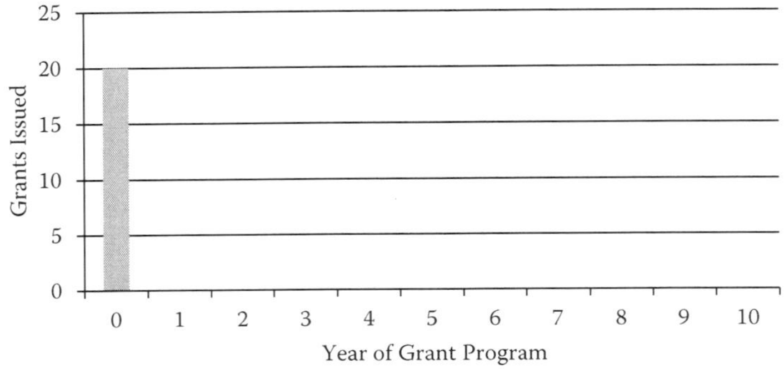

FIGURE 16.1
NEF grant program.

Figure 16.1 shows the impact of making twenty grants in the first year.

> Notice that at the end of the first year, there is no money left in the NEF account to finance future projects. A total of twenty projects were funded with its $100 million. That is it. If they choose grants, the game is over!

Subsidized Loans

Subsidized loans are loans that have an interest rate that is lower than the market rate. There is a cost to subsidizing loans. The cost of a subsidized loan is the present value of the difference between the payments on a market-rate loan of the same tenor and the subsidized loan. Thus, if the payments on a market-rate loan are $10,000 per year for 5 years and the payments on a subsidized loan are $6,000 for 5 years, then the cost of the subsidy is the present value of the difference in the two payments ($10,000 – $6,000 = $4,000) each year. The amount of the subsidy can be considered a grant.

The amount of a subsidy can be considered a grant because if the funds had been loaned out at market rates, the repayments to the NEF would be significantly higher. Thus, the lending institution—here, the NEF—"loses" money. Another way of looking at this is in terms of the growth of the NEF account. If it lends at a low rate, it receives lower repayments; which means that the NEF account grows much slower than it would have normally (i.e., if it had made market-rate loans). The shortfall in this growth is the value of the subsidy, which should be considered a grant.

With subsidized loans, the subsidy reduces the interest rate that borrowers pay on their loans. Lower interest rates result in lower periodic payments, leaving more cash available for the project, or otherwise enabling it to undertake a larger project than it could have without the lower, subsidized interest rate. This is very important. This is a strong plus for the concept of subsidized loans.

There will be circumstances where a community can only afford a $100,000 annual payment, but where a market-rate loan would require a $200,000 annual payment. In this case, if the project absolutely must go forward, the annual cost must be reduced. There are two ways of doing this: (1) an annual subsidy or (2) an upfront grant to reduce the amount of principal that must be borrowed and, hence, the annual payment.

As you know from the discussion of grants in Chapter 13, annual subsidies are an inefficient use of financial resources. Where borrowers cannot afford a market-rate loan for a given project, a grant should be employed to bring down the project cost. That is one of the few objectives for which grants are effective. However, the grant should not be an annual subsidy to reduce annual payments; rather, it should be a one-time, upfront grant—usually called a *buy-down*. The result is the same for the borrower—reduced costs—but it is much more efficient and cost saving for the lender.

Like grants, subsidized loans are used to facilitate projects that otherwise would not be financially feasible. To receive a subsidized loan, borrowers often must meet the requirements and conditions set by the government agency or private donor that is providing the subsidy. These loans have a higher efficiency than grants because part of the project's funding is paid back; however, there is still a capital loss to the NEF equal to the amount of the subsidy. As indicated above, the subsidy amount is calculated by subtracting the subsidized interest payments from current market interest rate payments, multiplying this difference by the outstanding balance of the loan each year, and then discounting each amount back to a present value.

> In our second simulation, the NEF managers decide to use the $100 million to make subsidized loans to finance all environmental projects. Loans are for a period of 5 years, repaid according to the *level principal method*. The interest earned on subsidized loans is assumed to be zero. All money that is received by the NEF as repayment of principal is reloaned in order to finance more projects. Because all projects require $5 million, there may be some money (less than $5 million) rolled over to the next year. The simulation occurs over a 10-year period. New loans are made each year from the annual principal repayments. Table 16.2 shows the results of using the $100 million to conduct a subsidized loan program.

In Table 16.2, notice how the same amount of seed capital ($100 million) provides financing for eighty-one projects when using subsidized loans as compared with only twenty projects financed by grants. Here, subsidized loans are funding 405% more projects than grants. So, subsidized loans are much more efficient than grants. But that is about it.

Figure 16.2 illustrates how many subsidized loans can be made with the same $100 million over a 10-year period.

TABLE 16.2
NEF Subsidized Loan Program

	Year										
	0	1	2	3	4	5	6	7	8	9	10
Opening Balance	100	0	0	4	2	1	2	1	2	0	4
Interest Earned	0	0	0	0	0	0	0	0	0	0	0
Principal Repaid	0	20	24	28	34	41	29	31	33	34	33
Total Funds Available	100	20	24	32	36	42	31	32	35	34	37
Amount lent	100	20	20	30	35	40	30	30	35	30	35
Ending Balance	0	0	4	2	1	2	1	2	0	4	2
Outstanding Loan Balance	100	100	96	98	99	98	99	98	100	96	98
New Loans Made	20	4	4	6	7	8	6	6	7	6	7
Total Loans Made	20	24	28	34	41	49	55	61	68	74	81

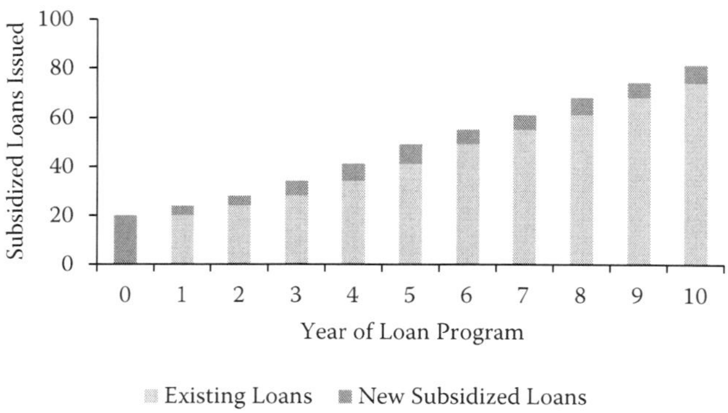

FIGURE 16.2
NEF subsidized loans issued.

Market-Rate Loans

An option for providing funding for more environmental projects than could be achieved with grants or subsidized loans is to provide loans at the market interest rate. Because market-rate loans do not provide a subsidy, they usually do not have conditions regarding the way that projects are carried out; in this sense, they are easier for borrowers to obtain than grants and subsidized loans. The higher interest rate (as compared with the subsidized loan) increases the total amount of the loan repayment, which results in higher periodic loan payments. These higher loan payments decrease the amount of available cash flow for the project.

Municipal Bonds

Before going to the market-rate loan simulation, let us discuss municipal bonds, which are a sub-category of market-rate loans. As you know, a bond has come to mean a loan (1) to a borrower with a well-known credit rating and (2) having highly standard documentation.

Two friends can make a $1,000 loan agreement on the back of an envelope. This does not work for cities or counties. Municipal bonds are loans to state or local governments, or their agencies. The lenders are most commonly the general public. The interest on most municipal bonds is exempt from both federal and state income taxation. This means that under normal circumstances,* the interest rates on such bonds are lower than the rates on US Treasury bonds or on high-quality corporate bonds.

Municipal bonds are usually either *general obligation* bonds or *revenue* bonds. With revenue bonds, specific municipal revenues (taxes and fees) are pledged to their repayment. With general obligation bonds, state and local governments promise to use 100% of their resources and powers to repay the debt.

Municipal bonds may be issued for any legitimate governmental purpose, including constructing schools, highways, water and sewer projects, electrical utilities, public housing, public hospitals, and other government facilities. In many cases, depending on individual state laws, the issuance of a bond (to borrow money) might require the permission of voters, in which case a local referendum is held.

Now, a municipal bond "issuance" is quite an elaborate event, as you might expect when a government borrows a large sum of money.† Let us take the City of Dallas as an example. If Dallas needed to borrow $100 million, they would first hire a financial advisor (FA), who would tell them the optimum structure (term, repayment schedule, etc.) for their loan. Then they would hire a bond counsel, who is a lawyer who specializes in these transactions. Both are hired through competitive bidding processes.

Then, when all is prepared and the FA thinks the time is right, Dallas offers the bonds for sale to the bidder (underwriter or syndicate of several underwriters) who offers them the lowest interest rate. The underwriters, then, immediately sell the bonds to you and me. The transaction between Dallas and the underwriters is called the *primary market*. The transaction between the underwriters and us is called the *secondary market*.

If governments borrow for five years or more, the borrowing is called a bond; if less than five years, it is called a note. There are several kinds of notes: Bond Anticipation Notes (BANs), Tax Anticipation Notes (TANs), and

---

* Not, however, in the continuing fall-out from the sub-prime mortgage recession that exists as of this writing, where Treasury rates are actually below comparable highly rated municipal bond rates.

† There are some modest variations in the way bonds are issued but they are, indeed, modest. And, at the growing behest of the Municipal Securities Rulemaking Board, these variations will become more and more modest.

Revenue Anticipation Notes (RANs). Some indication of their structure is implied in the title. A BAN is short-term borrowing, which is repaid when a bond, or long-term borrowing, is issued. A TAN is a short-term borrowing that is repaid when the government collects specific taxes. A RAN might be where a sewer authority borrows in order to pay for connecting a group of houses to a new sewer line for which it will charge connection fees. So, the RAN is repaid when the authority receives the connection fees.

All in all, the municipal bond market is a very large, safe business. There are currently about $3.7 trillion of municipal bonds outstanding. Before the subprime recession, there were about $400 billion of municipal bonds issued each year. The default rate on municipal bonds is 0.1%, according to Moody's Investors Service (henceforth "Moody's") and 0.29% according to Standard & Poor's.

So, municipal bonds are a form of market-rate loans. Now let us look at our comparative scenario for these market-rate loans.

> In Scenario 3, NEF managers now choose to provide all environmental projects financing with market-rate loans. The current market rate is 10% for loans of $5 million over a 5-year period. Payments are calculated based on the level principal method (as in the previous example). The results of a market-rate loan program are presented in Table 16.3.

Now that the NEF is no longer providing a subsidy to all the environmental projects that it is financing, many more projects get financed over the 10-year period. Both interest earned and principal repayments are recycled and re-loaned in order to provide financing for more projects, resulting in 145 projects being funded over the 10-year period. Market-rate loans provide a 725% increase in the number of projects financed over grants and an 80% increase in the number of projects financed over subsidized loans!

Figure 16.3 illustrates how many projects can be funded with market-rate loans.

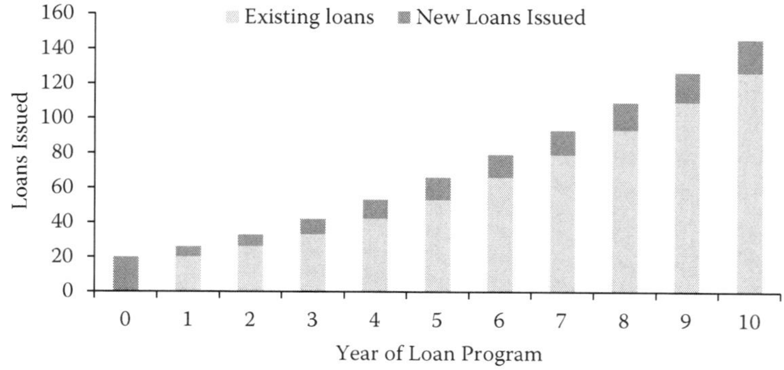

FIGURE 16.3
NEF market-rate loans issued.

TABLE 16.3
NEF Market-Rate Loan Program

	Year										
	0	**1**	**2**	**3**	**4**	**5**	**6**	**7**	**8**	**9**	**10**
Opening Balance	100	0	0	2	1.9	2	4.4	1	1.5	0.7	3.9
Interest Earned	0	10	11	11.9	13.1	14.4	15.6	17.5	19.2	21.2	23
Principal Repaid	0	20	26	33	42	53	46	53	60	67	73
Total Funds Available	100	30	37	46.9	57	69.4	66	71.5	80.7	88.9	99.9
Amount Lent	100	30	35	45	55	65	65	70	80	85	95
Ending Balance	0	0	2	1.9	2	4.4	1	1.5	0.7	3.9	4.9
Outstanding Loan Balance	100	110	119	131	144	156	175	192	212	230	252
New Loans Made	20	6	7	9	11	13	13	14	16	17	19
Total Loans Made	20	26	33	42	53	66	79	93	109	126	145

Loan Guaranties

Loan guaranties use the cash held by the NEF as collateral for guaranties of the repayment of loans made by private banks (or other lenders) to borrowers for environmental projects. In its role as guarantor, the NEF only needs to release money to lenders in the case of nonpayment by the borrowers. This structure allows the borrowers to negotiate favorable lending terms with the bank, usually resulting in below-market interest rates, longer repayment periods, and larger loan amounts. Borrowers are able to secure lower interest rates because the guaranty provided by the NEF reduces the credit risk that is assumed by the lending bank. Loan guaranties provide more benefits to borrowers as compared to borrowing at market rates, although the benefits are not as high as in the case of grants or subsidized loans. Generally, borrowers must comply with the terms and conditions that are established in the guaranty agreement, which are similar to those that are customary in market-rate loan covenants.

> The NEF managers have decided to allow existing banks (or other money-lending institutions) to provide the loan funds for the environmental projects. To ensure that the environmental projects are eligible for private financing and receive favorable treatment, the NEF will agree to guaranty the loans. Before we can roll out the impact of using our $100 million to provide loan guaranties, we need to add several new rules of the game.

Rules of the Game

So far, we have had six Rules of the Game. To recap, they are

1. Government contributes $100 million to the NEF
2. Project size: $5 million
3. Term: 5 years
4. Payment terms: Level principal method
5. Interest rate on subsidized loans = 0%
6. Interest rate on market-rate loans = 10%

To these six rules, we now add two more with the especially important leverage paradigm.

7. Interest rate on guaranty fund = 5%
8. Leverage Ratio Paradigm*:

Number of Loans	Coverage
0–20	100%
20–30	90%
30–40	80%
40–50	70%
50–60	60%
60–70	50%
70–80	40%
80–90	30%
90–100	20%
100+	10%

Now, Table 16.4 presents the results of an NEF program to use their $100 million to provide loan guaranties.

The bar graph in Figure 16.4 illustrates how many projects can be funded with the same $100 million by using loan guaranties.

With loan guaranties, the NEF can fund a total of 543 projects over a 10-year period! This amounts to 2,715% more projects than could have been funded by grants, 670% more projects than could have been funded with subsidized loans, and 374% more projects than could have been funded with a market-rate loan program.

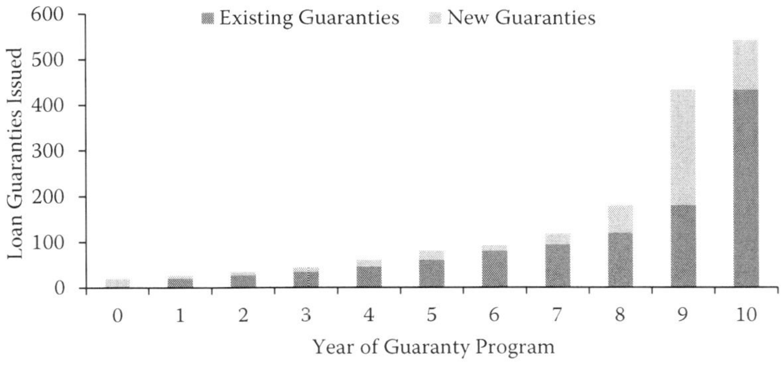

FIGURE 16.4
NEF loan guaranties issued.

* Estimating coverage between the minimum (>20) and the maximum (<100) loans is a totally inexact science; but what is presented certainly suffices for our illustrative purposes.

TABLE 16.4

NEF Loan Guaranty Program

	Year										
	0	**1**	**2**	**3**	**4**	**5**	**6**	**7**	**8**	**9**	**10**
Opening Fund Balance	100	100	105.5	111.4	117.5	124.1	131.2	138.9	147	155.8	166.2
Interest Earned	0	5	5.3	5.6	5.9	6.2	6.6	6.9	7.3	7.8	8.3
Guaranty Fees Earned	0	0.5	0.6	0.6	0.7	0.9	1.1	1.2	1.5	2.6	8.3
Closing Fund Balance	100	105.5	111.4	117.5	124.1	131.2	138.9	147	155.8	166.2	182.8
Leverage Ratio	100%	90%	90%	80%	70%	60%	60%	50%	30%	10%	10%
Maximum Guaranties	100	117	124	147	177	219	231	294	519	1662	1828
Begin Guaranties In Use	0	100	115	123	145	176	218	231	293	516	1659
Guaranties Released	0	20	27	33	44	58	57	63	82	132	372
Guaranties In Use	0	80	88	90	101	118	161	8 6	211	384	1287
New Guaranties Issued	100	35	35	55	75	100	70	125	305	1275	540
End Guaranties In Use	100	115	123	145	176	218	231	293	516	1659	1827
Guaranties Issued	20	7	7	11	15	20	14	25	61	255	108
Total Guaranties Issued	20	27	34	45	60	80	94	119	180	435	543

TABLE 16.5

Comparison of NEF Financing Schemes

	Year										
	0	1	2	3	4	5	6	7	8	9	10
Total Grants Made	20	0	0	0	0	0	0	0	0	0	0
Total Subsidized Loans Made	20	24	28	34	41	49	55	61	68	74	81
Total Market-Rate Loans Made	20	26	33	42	53	66	79	93	109	126	145
Total Guaranties Issued	20	27	34	45	60	80	94	119	180	435	543

Table 16.5 compares the effectiveness of these four financing techniques.

Let us turn these project numbers into project costs:

Grants	20	$ 100 million
Subsidized loans	81	$ 405 million
Market-rate loans	145	$ 725 million
Loan guaranties	543	$2.715 billion

Just look at that! $2.715 billion worth of environmental improvement projects! All for the same $100 million that bought you twenty grant projects!

Loan guaranty programs are so efficient because they incorporate *leverage.* Leverage, as you know, is the ability to increase the effect of the use of money. The same sum of money ($100 million) was used in each financing scheme. Under the subsidized loan program, there was modest leverage because the funds were paid and could be re-loaned for another project. Thus, the same amount of money could be used twice. This is leverage.

In the market-rate loan program, there was additional leverage because not only was the $100 million of principal repaid, but 10% interest was also paid each year. So, this increased the leverage.

With loan guaranties, even greater leverage occurs because of the principle of insurance. A guaranty is the same as an insurance policy.

A loan guaranty program insures "prompt and full payment" of debt. It works just like an insurance policy. With $100 million, a fund such as the NEF should easily be able to guaranty over $1 billion of projects at any one time because of the extreme unlikelihood that more than 10% of its projects would ever go into default at any one time. Thus, if the NEF were to guaranty commercial bank loans for $1 billion of projects, and 10% of them, or $100 million of them, were to default, the NEF could still make good on its guaranties by paying the banks holding the defaulted $100 million held in their reserve account.

Thus, from the point of view of a government wishing to finance environmental projects, the creation of a loan guaranty program would be, by far, the most effective use of its funds.

When you recall that the overarching goal of environmental finance is to provide the greatest amount of environmental benefits to the largest number of people at the lowest possible cost, then you can easily see that loan guaranties are the best possible means of financing environmental projects.

Grants are certainly the lowest possible cost. Subsidized loans also cost less but they fail to deliver the greatest amount of financial benefits to the largest number of people.

17

Cost/Benefit Analyses

For most people, the idea of conducting a cost/benefit analysis is something for economists and financiers to do. Right? Wrong! And here is an example.

The Rolling Stones are going back on tour. Ticket price: \$250. Tickets for self and significant other: \$500. Round-trip train to concert city: \$100 × 2 = \$200. Good dinner for two: \$200. Total evening: \$900. Do I like the Rolling Stones? Yes. Is this some special occasion? No. Do I like the Stones \$900 worth? No. So, are we going? No.

That is a cost/benefit analysis. And, of course, in reality, *everyone* is *constantly* conducting cost/benefit analyses, weighing the pros and cons of their options. And this process extends out to nearly every decision a person makes, however unconsciously. And so we enter the heart of economics, where people are totally rational and make their best decisions at what economists call the *margin*, comparing potential gain with opportunity cost. The *margin* simply means the next possible unit—*What would be the best way for me to spend the next hour of my time? What is the best use of \$900?* In other words, thinking at the margin brings cost/benefit analysis down to its most microscopic level. However, as the rest of us non-economists well know, people are often not rational or they do not have the luxury of making truly rational decisions because, frequently, they are forced to make decisions not on the margin but well beyond it. Being forced to make decisions so far in advance (and often without the best or complete information on all possible options) effectively bars a rational decision from being made as potential benefits and opportunity costs become more difficult to see and measure. This becomes an even more harrying situation when cost/benefit analyses are conducted between people of varying cultures, backgrounds, politics, and beliefs, as their matrices for cost/benefit may take conflicting factors into account that others may not find necessary or worthy of consideration in such a context.

Now, some of you may be asking: What *cultural* differences could impact a cost/benefit analysis?* After all, what does culture have to do with environmental finance? As it turns out: everything. Consider, for example, the construction of the Kinzua Dam. In the early 1960s, the Kennedy Administration broke one of the country's oldest treaties with the Seneca Nation in order to

* It is important to note, however, that cultural and other sorts of immeasurable concerns are not taken into account in official matrices of cost effectiveness.

construct the Kinzua Dam needed to protect a number of Pennsylvanian suburban neighborhoods from flooding as well as provide much-needed energy. Of course, in the process, they also forced Seneca Natives from their homes while also permanently disturbing and destroying Seneca burial grounds as the dam formed the deepest lake in Pennsylvania, the Allegheny Reservoir. In this situation, obvious cultural differences unveiled very different results from cost/benefit analyses that, more likely than not, took very similar things into account. And when the Seneca Nation took their case all the way to President Kennedy's desk, Kennedy was also forced to conduct a cost/benefit analysis regarding this issue of environmental health and energy supply, and found that, due to the benefits proposed through the construction of the dam combined with the Seneca Nation's lack of political clout in comparison to the voters of Pennsylvania and the burgeoning energy crisis, he favored the construction of the Kinzua Dam, thereby relocating the Seneca Nation yet again.

Thus, when it comes to the world of environmental finance, cost/benefit analysis, as you might imagine, is not a simple pros and cons list. Cost/benefit analyses must strip circumstances of cultural, political, and social ties from the issue at hand in order to reveal any actual, measurable cost-effective options. In other words, cost/benefit analyses related to environmental finance decisions are conducted at a significantly more intentional level with clear matrices laid out to guide decision making and measure benefit versus cost outside the realm of the subjective. These cost/benefit analyses are conducted in order to isolate *the project that requires the smallest amount of grant funds and that does the most environmental good for the largest number of people.*

Please read that last sentence again. There is one very important word in it: *grant*. Note that we are not trying to identify the project that costs the least; rather, we are looking for the project that requires the least amount of grant funds. This is very important, as you will see. So, when we actually repeat the mantra of the first principle of environmental finance—to provide the greatest environmental benefits for the largest number of people at the lowest possible cost—what we actually mean is *the lowest possible amount of grant funds*. There is a big difference here. If the costs are covered by loans, then those loan funds will be repaid and can eventually be reused for other projects. In the words of many popular environmental finance programs: *they revolve*. They can be used to pay for future projects. Grant funds, on the other hand, once spent, are gone—forever. So, it is the grant part of any project cost that must be minimized at all cost.

Now, the overwhelmingly important question that faces government financial policy makers is: *Which project receives a grant? And, how much should the grant be?* The answer to these questions is deceptively simple: "The project that requires the smallest amount of grant funds and does the most environmental good for the largest number of people." This answer can be embodied in a rubric that can be used as a rule for grantmakers.

For drinking water projects, the rubric can be stated as follows: "The highest priority for the use of limited grant funds is for projects that:

1. Reduce or eliminate the most severe waterborne disease
2. for the greatest number of people while
3. requiring the smallest amount of *grant*."

For wastewater projects, the rubric would be very similar, as follows: "The highest priority for the use of limited grant funds is for projects that:

1. Reduce or eliminate the largest quantity of
2. the most toxic pollutants while
3. requiring the smallest amount of *grant*."*

As is apparent, the above rubrics relate the cost (in terms of grant amount) to the benefits it produces. In short, to effectuate this rubric, a government environmental finance program manager must undertake two simple—albeit essential—steps before awarding any project grant funding: He/she must (1) perform a cost/benefit analysis and then (2) prioritize the results of that analysis. To achieve these goals, the public official must create two matrices to deal with the data necessary for a cost/benefit analysis. All of this can be accomplished in eleven simple steps. The first 5 steps relate to the benefit matrix.

Benefit Matrix

These requirements make up the "benefit" side of the cost/benefit analysis. We will use a contaminated wastewater scenario as an example.

1. When evaluating possible projects, categorize the environmental problems from bad to worst. When looking at wastewater problems, dysentery is bad but cholera and typhoid are much worse.
2. Identify where these problems are occurring. In what communities or locations?
3. Determine the frequency of the problem. How often are the outbreaks of waterborne disease occurring?

* Note that the rubrics above specifically say grants, not loans, or alternative funding sources, or anything else for that matter. Just grants. As you know, once a grant is made, the money is gone for good, unlike loans, where, once the loan is made, it must be repaid (with interest!) and the repayments can be recycled into new loans.

4. Determine the number of people who can potentially be affected. How many people live in the villages that are beset by frequent outbreaks of waterborne diseases?
5. Rank the projects where the greatest number of people are being affected most frequently by outbreaks of waterborne diseases.

You have just finished the benefit matrix.

The first matrix is thus constructed, consisting of (1) the ranking extent of waterborne diseases by severity, (2) the ranking of waterborne diseases by occurrence, and (3) the populations in areas of waterborne disease occurrence. This matrix then defines the first 2 elements of the following rubric: "places where the greatest number of people suffer most frequently from the most severe waterborne diseases."

Cost Matrix

The second matrix represents the factors encompassed in the "cost" side of the analysis. This matrix is more complicated and takes many more elements into account than its opposite, but they can easily be understood and feed into each other intuitively. These elements are as follows:

1. Determine the cost of a project to remediate the problem.
2. Identify all the financing programs that can be used to address the problem.
3. Calculate the Annual Debt Service Payments (ADSPs) on each of the different financing programs.
4. Determine how much ADSP the ratepayers can afford.
5. Convert the non-affordable part of the ADSP back into a principal amount, which is the amount of grant the project will need.
6. Rank the projects by least amount of grant needed.

You have just completed the cost matrix.

All that remains now in the cost/benefit analysis is to correlate both matrices. This correlation will define where the greatest need, occurring most frequently and affecting the largest number of people, can be met with the smallest amount of grant funds.

Case Study: Republic of Georgia Water Project

In the early 2000s, the US Environmental Protection Agency (EPA) asked a few colleagues and me to plan a safe drinking water program to be funded as part of the then-new Millennium Challenge Corporation (MCC) grant to the country of Georgia that the Georgians believed would be in the $30-million range.

At the very onset, there were two major issues to consider: (1) the health of the country's overall finances and (2) the state of their current tax system (in this case, specifically water tariffs). Thus, considering not only Georgia's high rate of poverty (with 34.3% urban, 37.1% rural, and 35.7% nationwide poverty levels in 2004 according to a 2006 study conducted by the IMF*), but also the fact that it was going to take several years before their water tariffs would reach full-cost recovery levels, we were able to quickly estimate a new, more realistic cost for the problem: over $200 million.† In other words, we were going to attack a $200-million project with only $30 million. So, grants were out of the question. We needed a program with leverage. We needed to leverage the $30 million into $200 million. But we also knew right from the start that we were going to require a great deal of grant funding in order to buy-down the project costs to affordable levels.

Hence, of the $30 million we were looking for, we calculated that we would need approximately $12 million to capitalize a revolving loan fund and then another $18 million in grants. Of course, realizing that the revolving loan funds would grow over time and that the grant money—once spent—would be gone forever (as another MCC grant seemed out of the question), we needed to devise the most productive use of the existent grant monies. All of this, as you might imagine, begged the deceptively simple question: What is the most productive use of this grant funding?

In order to help us begin to answer such a question, we turned to a series of cost/benefit analyses. But before I get ahead of myself, let me first explain how these conversations began in order to illustrate just how political a cost/benefit analysis can become. The Georgian Ministry of the Environment took the lead on this project and thus became responsible for identifying sustainable, efficient, and productive projects that would contribute to the Georgian government's safe drinking water initiatives. In order to do so, the Ministry conducted a series of analyses, including a Health Assessment, a Cost Assessment, an Ability to Pay Assessment, and a Cost/Benefit Analysis.

Their Cost/Benefit Analysis was to be based on "the number of expected beneficiaries (from the Health Assessment) and the expected costs (from the Cost Assessment)," and one was to be created for "each city or village water

* IMF Country Report No. 06/360, "Georgia: Poverty Reduction Strategy Paper Progress Report," International Monetary Fund, 2006: 7. <http://www.imf.org/external/pubs/ft/scr/2006/cr06360.pdf>

† The actual number was $295 million.

system whose supply demonstrates persistent contamination by waterborne pathogens." Given this analysis in combination with the Ability to Pay Assessment, the Fund would then be able to create a ranked list of cities with those with the highest potential benefit and lowest costs taking top priority. It is important to note, however, that the "lowest cost" for the Fund "is defined as those projects that require the least amount of grant funding."

The Ministry of the Environment officials then organized meetings with the Georgian Ministry of Health, the Centers for Disease Control, the "National Water Authority" (which is in charge of preparing cost estimates to maintain all the major water systems in the country), and the state "Social Investment Fund." From these meetings, we were able to work with the Ministry to develop plans and methods of analyzing projects with public health data to first determine benefit and then to compare said data with project cost estimates. As you may begin to notice, issues of diplomacy as well as those of social justice should be kept in the forefront of one's mind when conducting an environmental finance cost/benefit analysis. After all, the next step in the process was not simply the creation of a cost/benefit rubric, but an in-depth discussion with the Minister regarding the reforms necessary to ensure the financial sustainability of our projects. And, once we had come to a plan of action that all were amenable to—after months of planning, meeting, calculation, and presentations—the Minister was finally able to meet with President Edward Shevernadze of Georgia, the man who would then either authorize our project or send it back to the drawing board.

Fortunately, he authorized the Minister to prepare an MCC grant application for $50 million and to convene an interagency workgroup to further develop the program and application. After all of this, we returned yet again to Georgia to meet with the Minister and were finally able to lay out the following points of action:

1. To develop a transparent and well-documented list of projects
2. To design a new, government-controlled financial institution to receive the MCC grant, finance the projects, and act as the revolving fund
3. To prepare an MCC grant application
4. And, to marshal all of the pro-Georgia forces in the United States and Georgia to support the application

But, let me backtrack a bit now and take a moment to discuss just what I mean by a list of projects. Take, for example, the project of selecting cities. The decision of which cities to include in the larger program should be based on two primary factors: a cost/benefit analysis and the willingness of the local authorities to cooperate (that is, (1) undertake the necessary reforms and (2) undertake the project itself (in an open and transparent manner)). This marks only one of many overlapping and intersecting cost/benefit analyses needed in order to build up to the larger analysis aimed at answering our original question: What is the most productive use of this grant funding?

At the onset, we thought that the cost/benefit analysis of the project city selection should be based on two benefit factors and five cost factors. The two benefit factors are (1) the incidences of outbreaks of waterborne disease (or the failure of a water system to pass chemical and/or bacteriological tests) and (2) population data. As we got into our research, we recognized that there was a third factor: frequency of outbreak of waterborne disease. Some areas seemed to be more prone than others to such outbreaks. This needed to be factored into our considerations. Given this benefit data, the workgroup was then able to assemble an initial list of possible cities where the greatest benefits would ensue from a safe water program based on three factors: (1) the severity of waterborne disease, (2) the population affected by such outbreaks, and (3) the frequency of such outbreaks.

On the cost side of the equation, the workgroup—despite already having estimated costs for each project—had to then examine the list of possible cities and consider the specifics of local water tariffs, local subsidies, local costs for operating the water system, and the rate of collections. By analyzing these factors, the workgroup was able to determine just how much cash it would be possible to generate to pay the debt service on the project. From projected CADS (Cash Available for Debt Service), the workgroup was then able to also calculate the amount of a loan to the particular system. And, by subtracting the amount of the loan from the project cost, we could then determine the amount of grant necessary to actually build the project. In other words, the grant amount was—for us—the real cost of the project, because once the grant money was gone, no more projects were going to be possible without another massive infusion of grant funds, which, at that time, was by no means likely. So, when it came to conducting the cost/benefit analysis of the safe drinking water program as a whole, we were able to utilize not only the data already collected and synthesized, but many of the same essential steps and processes as well.

First, we developed the following rubric in order to define just what "most productive use" meant for us and the specific choices we faced: "The highest priority for the use of limited grant funds is for projects which:

1. Reduce or eliminate the most severe waterborne disease
2. affecting the greatest number of people, while
3. requiring the smallest amount of *grant*."

Notice again that the rubric specifies *grant* and not loan, not funding, not anything else. Just *grant*. And, as you might also have noticed, our rubric implies the creation of a pair of correlating matrices. There is a cost matrix. And there is a benefit matrix. Together, they constitute the cost/benefit analysis.

As identified earlier, the rubric requires that the country must have or otherwise develop a means of identifying and quantifying outbreaks of waterborne diseases. In this case, however, the Georgian government officials responsible for water and public health security had no such scientific means and were forced to instead operate through the use of an anecdotal system. Of course, as also illustrated before, in the absence of numbers, anecdotes suffice. In the land of the blind, the one-eyed man is king.

Again, as demonstrated before, the country must also have access to accurate population numbers. Because of much undocumented population mobility, the Georgian numbers were highly suspect.

This makes up the first of the two matrices, consisting of (1) the ranking of extant waterborne diseases by severity, (2) the ranking of extant waterborne diseases by occurrence, and (3) the population sizes in those areas of occurrence. This matrix thus defines the first 2 elements of the rubric: "places where the greatest number of people suffer most frequently from the most severe waterborne diseases." This, in other words, is the "benefit" matrix and thus the "benefit" side of the cost/benefit analysis.

Conversely, as you might have guessed, the second matrix is the cost side of the analysis as well as being the more complicated of the two. However, despite its inherent complexity, its elements can be easily dissected, explored, and explained. These elements include:

1. The determination of the cost of the project
2. The determination of the annual debt service by analyzing the rates and terms of loans that are available for the utility's project
3. The determination of how much debt can be supported by the utility's CADS
4. The calculation of the smallest amount of grant necessary to undertake the project (which can be so isolated by subtracting the maximum debt possible from the total project cost)

These four elements comprise the second matrix.

Once these two matrices were constructed, the cost/benefit from here on out was simple. From the benefit matrix, we selected the city with the largest populations suffering outbreaks of the most severe waterborne diseases most frequently. From the cost matrix, we selected the city that needed the least grant funds. So the highest priority city in our program was the one where the "most severe waterborne disease affecting the greatest number of people could be reduced or eliminated with the smallest amount of grant funding." This, in other words, is the total cost/benefit analysis.

These procedures were designed and utilized in our assessment of Georgia's water safety program for two primary reasons. First, they were meant to demonstrate for the Georgians how to construct a transparent project identification procedure—one that is free of politics, favoritism, and bribery. Second, because a data-driven cost/benefit analysis that does not consider emotional, cultural, or political costs and benefits will produce the clearest path for the most efficient and productive use of funds. Thus, even if the Georgians were to require 100% of the program to be funded by grants, the above cost/benefit analysis system would remain the most useful for bringing discipline to an otherwise chaotic and obscure project selection process. In other words, even if a city receives a 100% grant and still needs some operating subsidies, there will be three possible good outcomes from utilizing the above procedures: (1) it will, at least, put the city on the road to self-sufficiency; (2) it will bring discipline to the management of their system; and (3) it will assure the people of other cities—those who did not receive a 100% grant—that the project selection system was both fair and reasonable.

All of this is simply to say that, when conducting a cost/benefit analysis for environmental finance programs, it is not a matter of simply chalking up a list of pros and cons. Such cost/benefit analyses must hew to the financial bone of a situation by peeling away the layers of cultural, political, and social ties and biases in order to unveil the most cost-effective options. In other words, by constructing and examining the correlation of benefit and cost matrices, one may be able to determine *the project that requires the smallest amount of grant funds and that does the most environmental good for the largest number of people.*

18

Credit Enhancement

The words *credit enhancement* have a slightly smarmy ring to them, like *preowned*. People with good credit pay their bills fully and promptly. But what about those people or institutions that need credit enhancement? Why do they need their credit enhanced? What is wrong with them?

For us mortals, if we have a credit problem, we have to deal with it on our own. How do we enhance our credit? We don't. Unless you call paying higher interest rates credit enhancement, which in a way it is: The bank is taking extra payments to protect itself against your nonpayment, partial payment, or late payment.

Oddly enough, governments and other large institutions have the same problems. Sometimes the problems are transient and can be solved by a minor tax or rate increase. Sometimes, however, with both the average joe and with governments, the financial problems are systemic and deadly serious. In the late 1970s, the City of New York went bankrupt. As of this writing, the City of Detroit has declared bankruptcy, and Harrisburg, the capital of the Commonwealth of Pennsylvania, is teetering on the edge.

So, what is credit enhancement? Essentially, *credit enhancement* is a euphemism for reducing the risk of default and delinquency on debt. In general, credit enhancement is used on (non-federal) government debt as well as some major corporate debt.

Reducing the risk of default on public debt results in longer terms and lower interest rates. Longer terms and lower rates mean lower annual payments. Lower annual payments mean that more projects get done, and the ones that are done have a greater chance of success.

Take the case of a Bus Rapid Transport (BRT) project in a major international city. (BRT projects are climate change projects, as they can significantly reduce automobile traffic.) The money to repay debt issued to finance the project can only come from two sources: fares and subsidies from the city. If the Annual Debt Service Payments (ADSPs) are high, then the fares must be high or the subsidies must be high, or both. High fares mean lower ridership, which is shooting yourself in the foot because getting people out of automobiles and onto BRTs is how you reduce greenhouse gas emissions and retard climate change. By the same token (no pun intended), higher BRT subsidies mean that some other vital area of government, like public health, or education, or housing, is getting less money. High interest rates and short terms spell a lose/lose situation for climate change projects such as BRTs.

So the purpose of credit enhancement is to reduce the risk of default or delinquency on debt. And the goal of using credit enhancement is to drive down the ADSP on project debt.

There are six major strategies for enhancing credit or driving down the cost—or ADSP—of project debt. All these strategies involve removing—or at least making very remote—the risk of financial loss. If investors believe there is a high risk of loss, they will demand high interest rates to compensate for the risk. They will also only offer short terms to minimize the time in which something can go wrong. As we have said, short terms and high interest rates can doom climate change projects.

Here is one of the basic principles of credit enhancement. It is basically the principle of insurance. It begins with a concept that involves organizing financings into *tranches.* This is the French word *tranche,* meaning slice. The concept here is to slice a financing into pieces with ascending orders of risk of loss. This may sound complicated but it really is not. Let me explain.

Tranching

Let us say there are 1,000 apartment units in various suitable places. Now, the probability that all 1,000 tenants will make full and timely rental payments in any given month is close to zero. Out of 1,000 tenants, at least a few will be sick, dead, broke, forgetful, or gone. Conversely, however, the probability that all 1,000 tenants will default on their payment is also close to zero.

What is the probability that one tenant will not make his payment? Very high; close to 100%. What is the probability that five tenants will miss their payments? Again, very high; again, close to 100%. What is the probability that 100 tenants—that is 10% of the total—will miss their monthly rental payments? Very low; very low, indeed.

Now, let us say that you want to buy these apartments—all of them. The current owner will sell them—all of them—for $70 million. Let us say that the monthly rent for each apartment is $600. Let us also say that maintenance, taxes, and insurance are $1.2 million a year, or $100 per month per apartment.

Now, here is a new wrinkle: You are going to have all the tenants pay their rent directly to Trustee Bank. You will instruct Trustee Bank to take $100 out of every tenant's monthly payment and put it in an operating account for taxes, maintenance, and insurance. You then instruct Trustee Bank to put the balance, $500 per tenant per month, into an escrow account. That means that if every tenant paid his rent in full, you would have $500,000 a month in rental income in your escrow account. That is a total of $6,000,000 a year.

Now you want to finance the purchase of the buildings, so you go to the bank and tell them that you can pay $500,000 a month. You would like a

20-year mortgage and you have heard that the going interest rate is 6%. So, you ask for a $70 million mortgage.*

What do you think the banker would say? I think the banker would ask you a politely rhetorical question, something like, "And how much of a down payment are you planning on making?" Of course, if the buildings cost $70 million and you asked for a $70 million mortgage, you obviously did not plan on making *any* down payment. If the banker had agreed, you would have scored an infinite leverage deal. You would have bought the buildings entirely with other people's money. This does not happen . . . at least with banks.

Bankers like down payments. They like to know that their borrowers have their own skin in the game. One of the major reasons for this is, as we know, the possibility that all $500,000 will materialize—in other words, that all 1,000 tenants will pay promptly each month—is zero. The banker knows this too. So, he is not going to give you $70 million.

Now, let us say that you take the same deal to a smart investment banker. The investment banker will know, as we said above, that the probability that 900 tenants (or 90%) will pay their rent in full, on time is very, very high. But as the 901st rent check comes in, he knows that he is getting closer to the probability that at least a few of the tenants will either be late, or just won't pay their rent. So, he decides that instead of one bond, there will be two. Or, more correctly, one bond with two tranches: the "A Tranche," which will be $63,000,000 or 90% of the purchase price, and the "Z Tranche," which will be $7,000,000, or the other 10% of the purchase price. The A Tranche will be virtually risk free. The Z Tranche will bear 99.99% of the risk of loss. (By convention, no matter how many tranches there are in a financing, the investment bankers refer to the tranche that carries the lion's share of the risk of loss the "Z Tranche.")

So the investment banker issues a new instruction to the Trustee Bank: *put the first 900 tenant payments into a new account called "Escrow A."* Because the probability that at least 900 out of 1,000 tenants will pay is virtually 100%, the investment banker knows he can get a good interest rate on Tranche A. He can get a 5% interest rate. In addition, in the bond market he knows he can get a 25-year term.

Now, $500 per month times 900 tenants is $450,000 a month or $5.4 million per year. The debt service on Tranche A, which is $63,000,000 at 5% for 25 years, is $4.47 million. So, with $5.4 million in Escrow A, there is $930,000 more than is needed to pay the debt service on Tranche A.

The investment banker instructs the Trustee Bank to put the excess $930,000 *along with all additional rental income* into an Escrow Z account.

Now, if all of the 1,000 tenants pay on time, there will be $1,530,000 in Escrow Z. But the probability that every single tenant will do so is slim. This

---

* A $70-million debt for 20 years at a rate of 5.96% requires a monthly payment of $500,000. Call it 6%. Close enough.

means that the rate on Tranche Z will be much higher and the term will be much shorter.

So, the investment banker tests the market and learns that the Z Tranche can only get a 10-year term and will carry an interest rate of 15%.*

As such, the annual debt service payment on Tranche Z will be $1,394,764. Thus, on a good day (when every tenant makes a prompt and full payment), there will be $1,530,000 in Escrow Z of which only $1,394,764 will be needed for the debt service on Tranche Z. So, there will be a $135,236 surplus. On a good day, this will be the purchaser's profit. On a not so good day, this surplus will be used to protect Tranche Z investors from shortfalls in their annual debt service payments.

As you recall each tenant pays $600 a month or $7,200 a year. Dividing $135,236 by $7,200 will tell you that there can be 19 nonpayments *every month* out of 1,000; or about 2%, which is just about right. The Tranche Z investors get a modicum of protection. And, as long as the purchaser is left with $1, his Return on Investment is infinite since he did not invest one penny of his own money in the project. He bought the apartment complex entirely with other people's money.

People often refer to tranches as "smoke and mirrors." That's sour grapes. As you can plainly see, they work!

And so, as you can also see, we can create a credit enhancement mechanism by tranching. In our case, all of the risk is shifted to Tranche Z. Thus the presence of Tranche Z has the effect of credit enhancing Tranche A.

Self-Funded Reserves

The ADSP for a project can be significantly reduced by replacing the equity (or at least part of it) with a Self-Funded Reserve (SFR). Municipal revenue bonds in the United States are often structured in such a way. Here, the borrower puts up a reserve against its own default. In the United States, this is done singly. This means that if a water system issues a bond, they generally add to the principal one full year's debt service and place it into the hands of a bank trustee. This is to protect against transient problems. A major water main break is a good example. Here, the problem might be so severe and so acute that it eats up all the utility's cash reserves and the money it had put aside for its next ADSP. In such a case, when the trustee bank does not receive the system's payment on the due date, it dips into the reserve fund that had been set up. Theoretically, once the system gets its repairs done and puts its fiscal house back in order, it should replace the reserve funds that had been spent.

* The Z Tranche is an "equity" tranche. See Chapter 14.

In climate change funds, the situation would be a little different. The mechanism just described in the United States works for one project and one bond; but in the climate change world, it would work at the fund level, that is, a fund composed of a hundred or more projects.

Here each project would contribute 10% to a reserve fund; but, in this case, it would be a "common" reserve fund. In the US example above, the reserve protected against the borrower's own delinquency. In this new climate change example, *each borrower's reserve fund protects against the defaults of each and every one of the projects in the fund.* So, in our example, each of the 100 projects would contribute to a common (self-funded) reserve.

An SFR is, essentially, an over-borrowing that is placed into an escrow account to pay for defaults. Thus, if a project cost is $1,000, the borrower borrows $1,100 and puts the additional $100 into an escrow account. The self-funded reserve is fully amortized by the annual payments of the borrower. This raises the cost to the borrower but not nearly so much as does equity. If a borrower borrowed $1,100 and put the $100 into an escrow account—instead of using 10% equity—his annual payment would increase to $88.27,* but far less than the combined $97.22 cost of the $900 debt and $100 equity.

The table below is a comparison of ADSPs, of various levels of equity, and the corresponding costs of an equivalent SFR.

Debt	Equity/SFR	AP with Equity	AP with SFR	Difference
90%	10%	$97.22	$88.27	$10.03 (10.1%)
80%	20%	$114.19	$96.29	$18.07 (18.6%)
70%	30%	$131.17	$104.32	$23.31 (25.7%)

The first column is the percentage of debt in the financing.

The second column is the complementary amount of funding coming either from equity or from an SFR.

The third column is what the annual payment would be if equity were used instead of an SFR.

The fourth column is what the annual payment would be if an SFR were used instead of equity.

The fifth column shows what the difference would be between a financing with equity and an identical funding with an SFR. The percentage difference is also noted in parentheses.

It is clear that an SFR can replace equity in project financing at a much lower cost.

* Using the same 20-year term and 5% interest rate as in the above example.

There are two additional benefits to SFRs. SFRs provide debt investors additional protection against loss *over time*. SFRs are invested in interest-bearing accounts. They grow. Furthermore, as each loan repayment is made by the borrower, the outstanding balance on the debt declines. The combined effect of a growing reserve and a shrinking debt balance is a significant reduction in the investors' risk of loss over time.

Here is what the first 5 years of this phenomenon look like, compounding the reserve balance at a rate of 3% per year:

	Year				
	1	2	3	4	5
Bond balance	$1,067	$1,032	$995	$957	$916
Reserve balance	$103	$107	$109	$113	$116
Coverage ratio	9.66%	10.28%	10.98%	11.77%	12.65%

In the above example, after 10 years, the balance on the bond would be $682, while the reserve would have grown to $134, offering almost 20% coverage. After 15 years, the bond balance would be only $382, while the SFR would have grown to $156, providing over 40% coverage.

There are some political benefits in the SFR concept as well. If the 100 developing countries borrow an extra $1 billion on top of the $10 billion they need for their projects, and they default, then the money they lose is their own. There is a certain fundamental principle of justice here: If you ask for help, you must first be willing to help yourself. All investors feel more comfortable if a project sponsor (borrower) has its own skin in the game. The SFR is just that.

So, Strategy 2 is: *Replace equity with SFRs wherever possible.*

Second Loss Reserve

Strategy 3, much like an SFR, is addressed to the developed world, or the "Annex I countries" in the parlance of the United Nations Framework Convention on Climate Change (UNFCCC).* Because the developing countries are going to help themselves by putting $10 billion of their own skin into the game in the form of a common SFR, it is now time for the Annex I countries to do their equivalent.

They could to so in the form of *a leveraged 20% second loss guaranty fund.* What is this? "20%" means they would guaranty $2 billion of the $10 billion of annual climate change project debt; and "second loss" means only after

* The UNFCCC listed the most highly developed countries in its Annex I.

all $1 billion of the (first loss) SFR had been paid out to investors in respect of defaulted projects. So, all of the 10% first loss SFR would have to be paid out before any of the Annex I countries are called upon to pay investors one penny on their guaranties.

Leverage is the key to this scheme. Without leverage, the idea is nothing more than the usual plea for the rich to bail out the poor. The way leverage can be used in this context can best be described by reference to the North American Development Bank (NADBank), which was formed in 1993 as a joint venture between Mexico and the United States pursuant to the North American Free Trade Agreement (NAFTA). NADBank was capitalized with $3 billion, $1.5 billion each from the United States and Mexico. But the key to this structure is that the $3 billion is not cash: only 15% (or $450 million) is. The balance is *call capital,* which is a direct pledge of the respective governments to provide up to an additional $2.55 billion, if called upon, to meet any of NADBank's lawful obligations. So, Mexico and the United States have a bank with $3 billion of equity, but with only $450 million paid in. This is leverage: Discharging 100% of an obligation with 15% cash.

This is the way the *leveraged 20% second loss guaranty fund* can work: The 41 (rich) Annex I countries need only come up with $3 billion cash (15%) and $1.7 billion (85%) of call capital to fund a guaranty that will only be called upon after all of the $1-billion first loss SFR, provided by the 100 developing countries with projects, is used up. Having a total of $3 billion of cash and guaranties to protect them, investors should be more than willing to come up with the $10 billion needed to fund climate change projects at very reasonable interest rates and for very long terms.

A Role for the International Development Banks

There is a very definite role for the international development banks (IDBs) in financing renewable energy and climate change projects in developing countries.

Above, we suggested that the borrowing countries themselves set up a $1-billion SFR as the Z Tranche. What, in the best of all possible worlds, should happen is that one or more IDBs would purchase the $1-billion Z Tranche. Then developing country borrowers would still put up a $1-billion SFR, only it would take the second loss position, after the Z Tranche. So, the SFR would become the *second loss tranche.* Then the 20% loss reserve put up by the Annex I countries would become the *third loss tranche.* The next tranche would be the *A Tranche,* which should be totally risk free and, thus rated AAA, giving the entire financing the lowest rates and the longest terms.

Here is what this scheme would look like:

First Loss Tranche (Z)	$0–1 billion	IDBs
Second Loss Tranche (C)	$1–2 billion	Borrowers' SFR
Third Loss Tranche (B)	$2–4 billion	Annex I countries
Fourth Loss Tranche (A)	$4–10 billion	Retail investors

This is a role that has never been played by any of the IDBs. It is a very, very valuable role to play. It is long overdue that the IDBs begin taking the first loss positions, or Z Tranches, in order to obtain the most favorable terms and conditions for the financings of these valuable projects.

Tax Revenue Intercepts

Strategy 4 is another self-help strategy that could be adopted by developing countries wishing to fund climate change projects. Here is how tax revenue intercepts (TRIs) can reduce the risk of loss, thus lowering the borrowing costs on climate change projects.

An investor in an alternative energy or sustainable transport project in a developing country ultimately has an electricity ratepayer or a fare-paying bus passenger—as well as some local government subsidies—to rely on for repayment of debt. Investors do not want to stay up at night worrying about whether the ratepayer, or the passenger, or the local government will pay. They want to sleep soundly. They can do so if they know that there is another—solid—source of revenue that can be drawn upon, if need be. What could those sources of revenue be?

Unrelated tax receipts! These are credit support mechanisms that can be used to assuage investors' fears of nonpayment. It is important to note that these mechanisms do not replace fares and subsidies. They are not the source of repayments. Rather, these mechanisms are stand-bys, fallbacks, or safety nets, just in case the fare revenues are off or the subsidies are not enough. They serve as guaranties. Here are some examples.

In the late 1970s, the City of New York was bankrupt. Nonetheless, they needed to borrow money to undertake needed civic infrastructure projects. But no one would lend to them. At that time, there was a sales tax in New York State that was shared with local governments. The tax was 5% in those days, 3% of which went to the State and 2% to the City. The City collected all the taxes incurred in the City and forwarded the State's share to the State Treasurer.

To overcome the City's credit problem, the City and State enacted laws to provide that all sales tax revenues in the City would be paid into an account in the City's name at an international bank acting as a trustee. So, the City itself never legally touched this money. The bank forwarded the State's share to the State Treasurer. But the bank held the City's share for the benefit of its lenders. (As the City dutifully made each payment on its debt, the bank trustee would release the funds it then held into the City's general account.) And so, the City was then able to borrow money by pledging the hundreds of millions of dollars of its sales tax revenues—in the hands of an independent bank fiduciary—to its new lenders. So, the City's local sales tax revenues were *intercepted* for the benefit of its creditors.

As it turned out, the City never had to use any of these *intercepted* sales tax revenues to repay its debt. They were able to make all payments out of their other general revenues. The *intercept* was truly a stand-by, fallback mechanism.

This intercept mechanism is a powerful device that can be used to reduce the risk of nonpayment to lenders.

India has an even more powerful variation of the local tax intercept. This mechanism has been successfully used to support local water system debt in the State of Tamil Nadu. Here, certain tax revenues are *owned by the local governments*, but they are *collected by the State governments*. In this case, the local government and the State government entered into an agreement with the local government's lenders such that if the local government ever failed to make a timely loan repayment, then the lender could go to the State, which would *intercept* enough of the local government's tax revenues to make up the missed payment to the lender. A similar tax revenue sharing structure exists in Turkey.

In Mexico, certain tax revenues are legally shared by the federal and state governments. Because both the federal and state governments participate in the benefits of these monies, they are called *participaciones*. These *participaciones* are collected by the federal government. If a state borrows money, it, of course, promises to repay the money with interest from its general revenues. But in addition, the state may pledge that, in the event that the state fails to pay in the normal course of business, the lender may apply to the federal government, which is legally obligated to *intercept* the state's *participacion* of the tax and pay the state's lenders whatever they are owed. Standard & Poor's reported that, in 2009, 80% of all municipal bonds issued in Mexico were supported by *participacion* agreements—but no urban transport project bonds.

There is another variation of tax intercept that might be called a *super tax intercept* ("super" in the sense of "above"). Recall in the above example of the City of New York that the State of New York also owned 3% of the 5% sales tax revenue. This money belonged to the State, not the City. Nonetheless, if the State really wanted to assist the City, and the State did not want to issue its own guaranty (or was itself not creditworthy enough to issue a guaranty as in, for instance, the case of Argentina), then the State could have agreed that its share of the sales tax revenue could have been pledged to the repayment of the City's debt.

As noted above, these credit support mechanisms are not the source of debt repayment; rather, they are a stand-by or a fallback in case the borrower cannot make a debt payment. But they are exceedingly powerful.

They ease investors' fears of nonpayment to the point where they will accept lower interest rates and significantly longer terms on their investments. This results in lower project debt payments from local governments and breaks the upward spiral of risk and cost.

TRIs are not the only credit support mechanism that developing countries can use to lower debt costs. The same intercept concept can be used for budget devolutions. Where a central government establishes a permanent system of providing direct payments to local governments based on formulae, these payments can be pledged to, and thus intercepted by, creditors of the local government. These might be termed *devolution intercepts* (DIs).

TRIs can provide powerful support to debt, helping to reduce interest rates and lengthen terms. DIs and non-sovereign guaranties are not quite as good, but are definitely better than nothing.

Externally Funded Guaranties

Strategy 5 involves guaranties that are funded with cash; but unlike the case with SFRs, the cash comes from another source, generally a government program. So, although sovereign guaranties are much despised by developing countries, something short of a sovereign guaranty can be established by the guaranty of a central government financial institution, like a national development bank.

At $100+ billion, the US Clean Water State Revolving Fund (CWSRF) is the largest dedicated environmental finance program in the world. It is just such an externally funded source of guaranties. All 50 US states and the Commonwealth of Puerto Rico participate in CWSRF programs, which are managed at the state level. Since its inception in 1987, the US Congress has funded the CWSRF at various levels. These funds are appropriated to the US Environmental Protection Agency (EPA) and then distributed to the states under a congressionally approved formula. Under the Clean Water Act, in order to receive CWSRF funds, each state must appropriate $1 for every $5 it receives from the EPA.

The CWSRF is a loan fund. Loans only—not a penny of CWSRF funding can be used for grants.* The key to the CWSRF is that its funds *revolve*. First, they are loaned by the state to a community wastewater system. Then as the loan is repaid, the funds *revolve* back to the state. They are then reloaned again and *revolve* back to the community. And so on. The system is a closed

* Other than special funds that were appropriated to the CWSRF program by the American Recovery and Reinvestment Act (ARRA).

loop. Furthermore, in order to receive federal CWSRF funds from the EPA, each state is required to build a legal firewall between CWSRF money and all other funds owned or controlled by the state. This means that state CWSRF funds are inviolable; they cannot be used for any other purpose. They cannot be purloined by predatory governors or state legislatures when fiscal times are lean.

Today, the *size* of the CWSRF is $100+ billion, which means that $100+ billion of projects have been funded. The *net assets* of the CWSRF are $40+ billion. Net assets equal (1) the congressionally contributed funding, plus (2) the 1:5 matching funds appropriated by the states and (3) interest earned on funds that have been loaned. The difference between the program *size* ($100+ billion) and its *net assets* is called *leverage*. The CWSRF's leverage ratio is only about 2:1.

Standard & Poor's (S&P) has studied the default histories of municipal bonds for almost a century. The default history on bonds issued to fund public wastewater projects is approximately 0.04 (or 2,500:1). In January of 2011, S&P published a document requesting comment on, *inter alia*, a new total leverage ratio for anyone guarantying public wastewater project debt. S&P said that if the guarantor wanted to maintain its highest investment-grade credit rating (AAA), the maximum leverage ratio could be no higher than 75:1. With $40+ billion of net assets, at a 75:1 leverage ratio, the CWSRF has the potential to leverage $3 trillion. $3 *trillion*.

Why is the CWSRF important to a discussion of credit enhancement? Specifically because the 51 CWSRFs have the legal authority to issue guaranties. And, theoretically, their collective guaranties, up to a total of $3 trillion, would be rated AAA.

Now, what is most important to realize is that the guaranties offered by CWSRFs are not government guaranties in the legal sense. No, when a CWSRF issues a guaranty, the underlying documents specifically provide that the only source of the guaranty is the CWSRF's net assets. In other words, these CWSRF guaranties are not the legal promise of any government; rather, they are guaranties backed by enormous amounts of cash.

When you think of "externally funded guaranties" like these, think of a bank Letter of Credit (LoC). What is the bank's LoC backed with? The bank's equity; its own funds. Here in our example, the CWSRF guaranties are backed by each CWSRF's net assets. No legal promises; just cash.

There are several international examples of these "externally funded guaranties." The Philippines has its well-known Local Government Unit Guaranty Corporation (LGUGC). Furthermore, each of the Scandinavian countries has a variation of the CWSRF that credit enhances, or provides guaranties on, local government debt.

Legal Guaranties

A legal guaranty is an enforceable contract whereby one party promises to make full payment for the debt of a second party if the second party fails to make a full and timely payment on such debt. If you lend $1,000 to your neighbor, and the neighbor's rich uncle guaranties the debt in writing, and if your neighbor fails to make an agreed-upon payment at an agreed-upon time, you can take your written guaranty from the rich uncle into court and compel him to pay.

Another matter to consider about guaranties is the financial relationship between the guarantor and the debtor. If I walked into an investment banker's office and told him I wanted to issue a bond for $10 million, he would undoubtedly be both incredulous and dismissive. But if I handed him a letter from the US Secretary of the Treasury saying that the US government would guaranty my $10-million bond, he would be very impressed and get right to work. I would wind up with a long term and a very low interest rate on my bond because of the guaranty.

Now let us reverse the situation and say that I agreed to guaranty all of the US Treasury's outstanding debt. That is about $17 trillion. Do you think that would result in lower interest rates for Treasury bonds? I think not. So, the point here is that the value of any guarantor is directly related to the strength of his credit.

Legal guaranties are a sore subject with most governments. The reason is very straightforward. When a government, say, the City of New York, guaranties a debt, the amount of that debt is counted 100% against the credit of the guarantor, that is, the City.

When we say, "counted against the credit," we mean two things. First, if there is a constitutional debt limit, the entire amount is counted against the guarantying government's legal debt limit. So, when governments guaranty a debt, they are burning up their available borrowing power. They might as well borrow the money themselves and re-lend it.

Second, guarantying a debt counts against the credit of the guarantor from the perspective of the international credit rating agencies as well. This has two ramifications.

First, when credit agencies look at governments (or corporations), they assess their overall financial capability. In the case of a city, they look at the wealth of the community. They look at the tax rates. They look at the local government's budget. And they come to a conclusion as to how much that governmental unit can borrow with a certainty of full and timely repayment. If the government's outstanding debt is below that number, and the government and the community being otherwise in good order, then the government will be awarded a AAA rating. The less this is true, the lower the credit rating will be. Once a credit rating gets below investment grade, which is BBB–, the guaranties become pretty useless.

This brings us to our second point about legal guaranties and credit ratings: Many of the countries that have serious need of sustainability and renewable energy projects have execrable credit ratings.

First, the countries that, according to S&P, have AAA credit ratings include Australia, Canada, Denmark, Finland, Germany, Hong Kong, Liechtenstein, the Netherlands, Norway, Singapore, Sweden, Switzerland, and the United Kingdom. (Notice that the United States is not on this list.) S&P has issued credit ratings for 128 countries. Those having AAA ratings number 13, or 10% of the total.

On the other hand, there are some 55 out of these 128 countries that have non-investment-grade, or *speculative*, credit ratings, which are below BBB–. Bonds with non-investment-grade ratings are also called *junk* bonds.

We present the list of the 55 non-investment-grade-rated countries in Appendix A. Take a look. You will see some surprises. Greece and Portugal are not really surprises. Their credit woes have been in the news for the past several years. Venezuela and Argentina are both very wealthy countries but have, historically, had problems paying their bills. Then there are Georgia and Belarus, the only two of fifteen former republics of the Soviet Union to be on the junk list. Finally, there are Hungary, Jordan, and Turkey—all modern industrial states.

But, of course, it is the non-surprises that are of concern. These are the countries that need climate change and renewable energy projects. The problem is that income projections for such projects are sometimes less than credible in the international capital markets. As such, they could use a guaranty to boost their credit. But if the national government itself has a junk bond rating, its guaranty would be of little value. The end result is that the project will suffer with short borrowing terms and high interest rates, which means that fewer needed projects will get done.

Financial Guaranty Insurance

Strategy 6 is to use financial guaranty insurance to enhance your credit, if you can find it and if the math works out in your favor.

As noted in Chapter 11, before the sub-prime mortgage crisis, there were fifteen financial guaranty insurance companies that, in 2005, guarantied almost 60% of the about $400 billion of municipal bonds issued that year. In 2011, there was one company, and now there are two.

This entire industry was born in 1971 with the creation of the American Municipal Bond Assurance Company (AMBAC). This company insured municipal bonds only! The second entrant in the field was the Municipal Bond Insurance Association (MBIA) in 1973, which was founded by a consortium of five major insurance companies: Aetna, Fireman's Fund, Travelers, Cigna, and Continental. MBIA also insured only municipal bonds.

This industry, composed of only two companies, churned along, making considerable profits until the mid-1980s when the first signs of trouble began to appear: *competition*.

From the mid-1980s through the 1990s, this little industry swelled from two companies to fifteen. Word got out about how much profit AMBAC and MBIA were making, and other investors flocked to join the field.

With fifteen companies now in the market, the second sign of trouble appeared: *There were now fifteen companies competing for business*. The natural thing to do, especially for the new guys trying to get a toehold in the business, was to cut premium rates. As premium rate cutting got worse, profits plummeted and investors became disillusioned.

And now the third sign of trouble appeared: *collateralized debt obligations (CDOs)*. Frantic over falling profits, one by one the companies were lured away from insuring traditional municipal bonds, where there was too much competition fighting for too few deals, into the heady realm of CDOs. These CDOs housed sub-prime mortgage obligations. There was no money to be made in insuring traditional municipal bonds, but there were plenty of profits to be made insuring CDOs. So, the old municipal bond insurance industry, that now called itself the financial guaranty insurance industry, plunged headlong into the uncharted waters of insuring CDOs.

To say they made a mistake is one of the great understatements of the century. The financial guaranty insurance companies totally miscalculated the risks inherent in these CDOs with their payloads of sub-prime mortgages. But the industry was not alone. Even the rating agencies miscalculated the risks and continued to affirm the AA and AAA ratings of these financial guaranty insurance companies, even as their risk portfolios became more and more bloated by the sub-prime risks. Finally, the entire empire collapsed. As the 100+% loan-to-value sub-prime mortgages came up for rate adjustments, people began to realize that there were thousands of homeowners who could not afford the new, higher rates. When they could not pay their mortgage that had been sold into a CDO by the originating bank, the CDOs started defaulting on *their* payments. When this happened, investors went to the financial guaranty insurers. The financial guaranty insurance companies had nowhere near enough reserves to pay such massive losses. And so, the financial guaranty insurance industry collapsed.

But, like the phoenix that rises from its own ashes, the industry will be back. There was one company in 2011. Another opened its doors last year. And at least one rating agency believes that there will be more in the next few years.

So, should you use municipal bond insurance to enhance the credit of your transaction? If it works for you, yes. What does this mean? It means that if the bond insurance will lower your interest rate more than enough to cover the cost of the insurance premium, then you should use it.

As explained in Chapter 11, because of the checkered history of bond insurers, their insured bonds sometimes do not trade as well as an uninsured bond with the same rating. In other words, if City A, with a AA credit rating issues a bond, and City B, with only a BBB rating but which buys insurance from a AA-rated insurer, issues an identical bond, you will often see City A's bond trading at a better rate than City B's.

So, the appropriate way to deal with bond insurance is to determine at what rates the two bond insurers bonds are trading. Determine where your bonds would trade without insurance. Then request a quote from the two insurers and determine if the lower payments because of the higher credit rating will more than offset the cost of the insurance.

In general, the rule is that the insurance premium should eat up just under half of the spread between what your bonds would trade at with the insurance versus without the insurance. In other words, as illustrated in Chapter 11, if your bonds would sell at 5% without the insurance and at 4.5% with the insurance, then the premium should cost you about 0.2%, which means your all-in rate with the insurance would be 4.7%. So you would save the difference between 5% and 4.7%, or 0.3%.

So, *credit enhancement* is not just some slick Wall Street terminology. Rather, it is a term that covers methods and mechanisms to lower your interest rate (and possibly lengthen your term) by improving the credit rating on your project.

Whenever you are involved in environmental or sustainability projects or renewable energy projects, your goal will be to drive down the cost of financing them. As such, you should consider any of these credit enhancement strategies that will help.

- First and foremost, as has been expressed several times in this book, avoid using equity if at all possible. It is so expensive.
- Second, divide your financing into tranches.
- Third, get an IDB (or a state agency such as an SRF) to purchase the first loss layer, or Z Tranche.
- Fourth, when there are multiple projects being financed together, have each borrower contribute to an SFR, which would constitute a "risk-proximate" layer like the Z Tranche, or right above it.
- Fifth, if working internationally, ask developing countries to fund another risk-proximate layer.
- Sixth, look for external guaranty or debt insurance funds.
- Seventh, look for a credit-worthy government willing to guaranty the financing.
- Eighth, check out whether financial guaranty insurance is available and whether it is cost effective to buy it.

Always remember that our overarching goal is to provide the greatest environmental benefit to the largest number of people at the lowest possible cost. Credit enhancement is a means of achieving the lowest possible cost.

19

Tariffs

Efficient tariffs reduce the cost of providing environmental utility services to customers. Lower rates mean that more environmental improvement projects will be undertaken. This chapter will discuss the various types of tariffs and how they can help achieve the goal of lower-cost environmental projects and services.

Characteristics of Good Tariffs

The following are some of the characteristics of well-considered tariffs for water and wastewater systems:

- *Revenue sufficiency/cost recovery*: Tariffs produce stable revenue that is equal to the financial cost of supplying the water or wastewater service.
- *Economic efficiency*: Water prices signal to consumers the financial, environmental, and other costs that their decisions to use water impose on the rest of the system and on the economy.
- *Fairness*: Equals must be treated equally; in other words, the prices charged to customers are equal to the costs imposed on the system by those customers.
- *Resource conservation*: Pricing decisions should not promote the unwise use of water resources.
- *Net revenue stability*: Prices should allow the utility to have sufficient income to meet its operating costs, even when quantities demanded are below a normal level.
- *Transparency*: Pricing structures should be able to be understood by every consumer, in order for the consumer to respond accurately when deciding how much water to consume.
- *Ease of implementation*: Pricing structure should not impose significant administrative costs on the utility.
- *Affordability:* The prices charged to customers should be within a standard limit of affordability.

The objectives listed above provide a guideline for utility managers to follow when deciding on water tariffs; however, not all of the objectives can be met at the same time. This chapter will focus on two of the objectives that are most directly related to the financial stability of the utility: cost recovery and economic efficiency. The objectives of equity and affordability have been discussed elsewhere in this book.

Full Cost Recovery Tariffs

When setting the tariff, utility managers must understand that they must at least be able to recover the cost of operating the water system from the revenues earned. If the utility did not have to provide for the cost of building a new project (i.e., the project costs), then the tariff must be set at the level where it will equal the amount of total cash expenses. One method for designing a tariff is to begin with the operating budget (equal to the total cash expenses) and divide that amount by the total number of households that will purchase the water service; the result is the per-household tariff required for the tariff to recoup the costs of operating the system.

It is highly unlikely that a utility can rely on grant funding for 100% of its project costs; it is probable that the utility will have to pay at least a portion of the project costs. The more project costs that are paid for by the utility, the higher the household water tariffs, as the utility must pass on the costs of providing services to the customers that use water. A helpful exercise will be to determine what the tariff must be for a community that is in need of a project to modernize its water system, so that the full cost of water delivery is recouped by the utility. This exercise is considered for different scenarios: 100% of the project funded by an outside grant; 75% of the project funded by a grant and 25% funded with a loan; 50% of the project funded by a grant and 50% by a loan; and 25% funded by a grant and 75% by a loan. Under these different scenarios, the important factors will be the term of the loan and the interest rate. Because income must equal the expenses (operating plus project costs), it is important to consider the effect of rate and term on the household tariff. It is possible to reduce expenses just as it is possible to increase income (i.e., the tariff) so that the tariff does not become unaffordable.

> In the following example, assume that your utility serves 1,000 households and your current expenses* of $10,000 per year are covered completely by a $10-per-household tariff. In order to modernize the water system and to repair pressing problems, a project is proposed that will cost $1,500 plus an increase in annual operating expenses of $550. Below

* Expenses include energy, labor, chemicals, and administration costs.

are a few possible effects on the tariff corresponding to different project financing options.

Financing Option	Annual Operating Cost ($)	Annual Project Cost[a] ($)	Total Annual Cost ($)	Full Cost Recovery Tariff ($ per HH)	Percent Change in Tariff
100% Grant	10,550	0	10,550	10.55	5.5
75% Grant, 25% Loan (5 yr, 10%)	10,550	112.5	10,662.50	10.66	6.63
75% Grant, 25% Loan (10 yr, 5%)	10,550	56.25	10,606.25	10.61	6.06
50% Grant, 50% Loan (5yr, 10%)	10,550	225	10,775	10.78	7.75
50% Grant, 50% Loan (10 yr, 5%)	10,550	112.50	10,662.50	10.66	6.63
25% Grant, 75% Loan (5 yr, 10%)	10,550	337.50	10,887.50	10.89	8.88
25% Grant, 75% Loan (10 yr, 5%)	10,550	168.75	10,718.75	10.72	7.19
100% Loan (5yr, 10%)	10,550	450	11,000	11.00	10.0
100% Loan (10yr, 5%)	10,550	225	10,775	10.78	7.75

[a] Using a level principle payment schedule, the total annual payment decreases annually; in this example, the annual project cost is equal to the first year's payment, and therefore subsequent year payments will be lower.

As expected, the more that a utility must pay for the project costs in a full cost recovery tariff system, the larger the impact on the household tariff. As shown above, a longer-term loan at a much lower interest rate has a smaller impact on the costs (and therefore tariff) than does a shorter-term loan.

Tariff Design Options

Given the requirement that utilities use full cost recovery methods as described above, there are a variety of ways that a utility can design the tariff scheme. To demonstrate the unique features of each tariff design option, the following example will be applied to each option:

A public utility has 100 customers that purchase the following amounts of water each month:

- 25 customers consume 500 cubic meters of water.
- 50 customers consume 1,000 cubic meters of water.
- 25 customers consume 1,500 cubic meters of water.

Fixed versus Volumetric Charges

Fixed Charges

The first way that we will distinguish tariffs is by those that have a fixed charge and those that have a volumetric charge. A fixed charge is one where the amount of money charged to a customer is the same each month and is independent of the amount of water that customer consumes.

In places where there are no water meters, volumetric charges are not a viable option because there is no way to accurately measure how much water each customer consumes. In this situation, a *single-part tariff, or fixed-charge tariff* using equal monthly charges is the only available option.

When setting the fixed charge, or fee, for all customers, the utility must first determine its own costs for delivering the amount of water demanded by all customers, because the total revenue that the utility will receive must cover the utility's total costs plus provide funds for maintenance and repairs. A utility may define different "classes" of customers, so that residential customers are charged different (usually less) fixed rates than business customers, as businesses tend to consume more water than residential users.

Fixed charges, when used alone, have two major drawbacks. First, consumers are given no incentive to economize on water use because each additional gallon of water comes free of charge. Second, as water use increases, as will certainly happen with a growing population and a developing economy, the utility's ability to recover its costs by the fixed charge will diminish as the costs of meeting growing needs increase.

In our example, all 100 customers are charged the same fee (in this case, it is set at $20 each month), even though they consume different amounts of water.

Consumption Level	Total Charge
500 m^3	$20
1,000 m^3	$20
1,500 m^3	$20

Volumetric Charges

A utility that has water meters on all or most household connections is able to monitor the amount of water that each consumer (i.e., household) uses and is thus able to charge customers according to the amount of water consumed each month. When setting this price, the utility should attempt to set the volumetric charge at the *marginal cost* per unit of water provided to each consumer.

The marginal cost estimate must incorporate administrative costs as well as the costs associated with each additional volume of water provided, because in a single-part tariff, there is no additional fixed charge added to consumers' bills to cover the administrative costs.

To calculate the total charge for each customer group (in this case, groups are separated by the amount of water consumed), multiply the amount of water consumed by the price per cubic meter of water. In this case, we have a uniform volumetric tariff structure where the price per unit of water is the same regardless of how much water is purchased in total by each customer.

In this example, a *uniform volumetric charge* is used, where each consumer group is charged the same volumetric price. The next section discusses other forms of volumetric charges.

The volumetric charge example is different from the fixed-charge example in that the charges now depend on the amount of water consumed, so each customer group has a different total bill for water. The total charges are calculated as follows using a uniform volumetric charge of $0.02 per cubic meter:*

Consumption Level	Total Charge Calculation	Total Charge
500 m^3	500 × $0.02	$10
1,000 m^3	1000 × $0.02	$20
1,500 m^3	1500 × $0.02	$30

Multipart Tariff

A tariff can be made up of either a fixed or a volumetric charge, as explained above, or it can have both, which would be a multipart tariff. In the case of a multipart tariff, the fixed charge element of the water bill tends to cover infrastructural costs (e.g., the cost of installing and reading water meters and billing customers) incurred by the utility in order for it to be able to deliver water to each customer, and the volumetric charge amount reflects the "per-unit" cost of delivering water to the consumer.

The fixed charge is in addition to the price of delivering the amount of water demanded by that customer (i.e., the volumetric charge) because it reflects the costs incurred by the water utility to provide services to customers regardless of the level of water consumption.

In our example, the fixed charge will be $10 per connection (i.e., household), and the uniform volumetric charge will be $0.02 per cubic meter of water consumed.

Consumption Level	Total Charge Calculation	Total Charge
500 m^3	$10 + 500 × $0.02	$20
1,000 m^3	$10 + 1000 × $0.02	$30
1,500 m^3	$10 + 1500 × $0.02	$40

Notice how the total charges are all simply $10 higher than the respective charges for the single-part uniform volumetric charge.

* A cubic meter is about 268 gallons.

Types of Volumetric Charges

There are a small number of accepted methods for designing volumetric charges that must be considered when setting tariffs, so that the utility is able to select a tariff that will achieve full recovery of the costs to deliver water service, while trying to achieve the other goals of equity, affordability, and economic efficiency. The four volumetric pricing design options, in addition to the uniform volumetric charge that was described above, are: increasing block tariff, decreasing block tariff, seasonal pricing, and zonal pricing.

Increasing Block Tariff (IBT)

In theory, IBT can achieve three objectives simultaneously:

1. Promote affordability by providing the poor with affordable access to a "subsistence block" of water
2. Achieve efficiency by confronting consumers in the highest price block with the marginal cost of using water
3. Raise sufficient revenues to recover costs

In practice, however, IBTs often fail to meet any of the three objectives mentioned above, in part because they tend to be poorly designed. Many IBTs fail to reach cost recovery and economic efficiency objectives, usually because the upper consumption blocks are not priced at sufficiently high levels and/or because the first subsidized block is so large that almost all residential consumers never consume beyond this level.

When used in a multipart tariff, the increasing block pricing scheme affects only the volumetric part of the total tariff; the fixed charge remains the same over all customer groups. Calculating the total charge for each customer group using an increasing (or a decreasing) block tariff first requires delineating the number of blocks and the quantity of water allowed in each block. Customers will pay the fixed charge plus the sum of the products of the amount of water consumed in each block multiplied by the per-unit price of water in each block.

In our example, the three blocks will be as follows:

- *Block 1: $0.01 per cubic meter for first 500 cubic meters of water consumed*
- *Block 2: $0.02 per cubic meter for amounts between 501 cubic meters and 1,000 cubic meters of water consumed*
- *Block 3: $0.03 per cubic meter for amounts over 1,000 cubic meters of water consumed*

Consumption Level	Total Charge Calculation	Total Charge
500 m^3	$10 + (500 × $0.01)	$15
1,000 m^3	$10 + [(500 × $0.01) + (500 × $0.02)]	$25
1,500 m^3	$10 + [(500 × $0.01) + (500 × $0.02) + (500 × $0.03)]	$40

Notice how the incremental increase of 500 cubic meters of water resulted in an exponentially higher total charge for the third block than for the second block.

The next tariff design option to be discussed is the decreasing block tariff (DBT).

Decreasing Block Tariff (DBT)

A decreasing block tariff (DBT) is the opposite of an IBT. With a DBT, consumers face a high volumetric charge up to the specified quantity of the first block. Then, any water consumed beyond this level, and up to the next block, is charged at a lower rate, and so on for as many blocks as the tariff utilizes.

The DBT structure was designed to reflect the fact that when raw water supplies are abundant, large industrial customers often incur lower average costs because they enable the utility to capture economies of scale in water source development, transmission, and treatment. This tariff design has gradually fallen out of favor, in part because marginal costs, properly defined, are now relatively high in many parts of the world, and there is thus increased interest in promoting water conservation by the largest customers. The decreasing block tariff structure is also often politically unattractive because it results in high-volume users paying lower average water prices.

In the decreasing block rate example, we will continue to use a fixed rate of $10 and the block delineations; however, the per unit price of water is now:

- *Block 1: $0.03 per cubic meter for first 500 cubic meters of water consumed*
- *Block 2: $0.02 per cubic meter for amounts between 501 cubic meters and 1,000 cubic meters of water consumed*
- *Block 3: $0.01 per cubic meter for amounts over 1,000 cubic meters of water consumed*

Consumption Level	Total Charge Calculation	Total Charge
500 m^3	\$10 + (500 × \$0.03)	\$25
1,000 m^3	\$10 + [(500 × \$0.03) + (500 × \$0.02)]	\$35
1,500 m^3	\$10 + [(500 × \$0.03) + (500 × \$0.02) + (500 × \$0.01)]	\$40

Notice how the total charge for the first two groups of users is now $10 higher than with the IBT, whereas the total charge for the largest water user did not change.

Seasonal Pricing

In some circumstances, the marginal cost of supplying water to customers may vary by season. In such cases, water tariffs can be used to signal customers that the costs of water supply are not constant across the seasons. Summer water use tends to be much higher for households that have gardens and any other high water uses; the increase in water use is usually found to be in outdoor water use, while indoor water use tends to remain constant throughout the year.

When the marginal cost to provide water services changes according to the seasons, then utilities can charge higher prices during the more costly seasons (usually summer) and lower prices during the less expensive seasons (usually winter).

In this example, we will return to using the uniform volumetric tariff that charges the same per-unit amount for water regardless of the amount of water consumed so that we can focus on the effect of a higher price for the summer season than for the winter season. We will assume that the amount of water consumed at all levels is constant during the year. The fixed charge is $10 and the seasonal prices are as follows:

- *Winter: $0.01 per cubic meter of water consumed*
- *Summer: $0.02 per cubic meter of water consumed*

Consumption Level

Summer Level	Total Charge Calculation	Total Charge
500 m^3	$10 + (500 × $0.02)	$20
1,000 m^3	$10 + (1000 × $0.02)	$30
1,500 m^3	$10 + (1500 × $0.02)	$40

Consumption Level

Winter Level	Total Charge Calculation	Total Charge
500 m^3	$10 + (500 × $0.01)	$15
1,000 m^3	$10 + (1000 × $0.01)	$20
1,500 m^3	$10 + (1500 × $0.01)	$25

A seasonal and/or zonal pricing (described below) scheme can be added to any of the three pricing methods above (uniform, IBT, and DBT), as these schemes are simply adjustments to the chosen pricing scheme based on the time of the year (seasonal pricing) and the location of customers (zonal pricing).

Zonal Pricing

Zonal pricing can occur when the marginal cost to provide water service varies according to the location of the customer. In this case, utilities can charge higher volumetric rates to the customers who live in an area that is more costly to serve, and lower volumetric rates to those who live in areas that are cheaper to serve.

It may cost the water utility more to deliver water to outlying communities due, for example, to higher elevations and increased pumping costs. Zonal prices can be used to ensure that users receive the economic signal that living in such areas involves substantially higher water supply costs. However, this type of special tariff is only appropriate if the costs to serve the area are significantly higher than for the rest of the community—in fact, costs vary among all users, and a practical tariff always reflects average costs to some degree.

In the zonal pricing example, we will continue to use a fixed charge of $10 and a uniform volumetric charge with variations in the per-unit charge depending on the zone in which consumers reside:

- *Zone 1: $0.02 per cubic meter of water consumed*
- *Zone 2: $0.03 per cubic meter of water consumed*

Consumption Level Zone 1	Total Charge Calculation	Total Charge
500 m^3	$10 + (500 × $0.02)	$20
1,000 m^3	$10 + (1000 × $0.02)	$30
1,500 m^3	$10 + (1500 × $0.02)	$40

Consumption Level Zone 2	Total Charge Calculation	Total Charge
500 m^3	$10 + (500 × $0.03)	$25
1,000 m^3	$10 + (1000 × $0.03)	$40
1,500 m^3	$10 + (1500 × $0.03)	$55

The goals of full cost recovery and economic efficiency are achieved when utilities charge customers according to the cost that each customer imposes on the system. If the cost to deliver water is the same for all of the utility's customers, then a uniform volumetric charge that is set equal to the cost of water delivery is the most efficient. The DBT is used when the cost to deliver large amounts of water is cheaper than for small amounts of water. When the marginal cost of water is high, IBT is usually chosen over a decreasing block tariff because it charges higher rates to customers who demand higher levels of water.

Regulation of Water Tariffs

There are many different types of utility systems throughout the world, including public agencies (local or national), government corporations, private but regulated companies, and private unregulated companies. These tend to be larger utility systems, while there are also small systems (usually rural) that are governed by local water committees. The size and legal status (i.e., public or private) of the utility system are important indicators of whether or not the utility system will need public regulation of the prices charged to customers.

- Community (Small) Water Systems are self-regulating, as the community comes together to determine how much money to charge each household for access to the community system.
- Medium and Large Urban Water Systems may require price regulations to protect the public from unfair pricing.

Public (i.e., governmental) water utilities set prices at levels that benefit society the most; this often means that prices are set so that the utility can recoup the costs of providing the service plus a small amount to cover future repairs and maintenance. Another requirement for public utilities is that all or almost all of the local population is covered by the water service. Prices of water that is delivered by private water companies in the absence of price regulation tend to be much higher because such companies are often *natural monopolies*.

The goal of a natural monopoly is often to maximize its own profits, which is easily accomplished because of its market power. Therefore, the monopoly will charge the highest price possible for any amount of water it delivers. This price is likely going to be too high for a significant part of the market to afford, which leaves the poorest people without this necessary service. Enforceable regulation of private water companies can correct these market failures.

Types of Regulation

There are two general types of regulation: *economic* and *social*. Economic regulation generally includes any policy or rule that governs the prices set by private companies, and the types and amount of goods or services supplied by those companies. Social regulation includes all other factors that are not covered by economic regulation; with respect to water utilities, this would include water quality standards. Both types require that an independent governmental agency be responsible for monitoring the prices charged by the utility and the quality of the water delivered by the utility; it is not expected that one agency should take on both roles, although this might be necessary in a small water system. The important aspect of regulation that is often lacking in many developing countries is the ability of regulating agencies to enforce their decisions. Enforcement can only occur where there is a legal basis for enforcement, and sanctions for noncompliance must be of enough value to prevent the utility from engaging in illegal or inappropriate activities.

The choice of how to regulate water prices must be made by the government with input from the utility, consumer advocates, and other interested parties. Regulating agencies have many available methods for deciding the appropriate price for the utilities to charge customers; however, the important rule is that utilities must be able to recoup their necessary costs of providing the service and be allowed to retain a small amount above their costs in order to save for future repairs and maintenance. Annual adjustments should be allowed, such as an annual increase equal to the rate of inflation, without requiring any extensive appeal to the regulatory agency.

Independent Financial and Management Audit

Monitoring of utility pricing and water quality is the most important role of the regulating agency. The utilities should not be allowed to report their prices and water quality measures to the regulators without an independent monitor to verify the accuracy of the data; this is the only way to ensure that society's interests are protected. The regulatory agency and the utility should agree upon a third party—a truly independent group of monitors—to conduct the annual economic and water quality surveys of the utility. A third party is most likely to provide an objective analysis that will benefit both the regulator and the utility.

Tariffs are an important piece of the puzzle of lowering the cost of environmental improvements. Appropriately designed tariffs can reduce the cost burden on consumers. To the extent these burdens are lower, there is more room for financing new projects to even further improve the quality of the environment.

20

Climate Change and Renewable Energy

We are going to look at climate and renewable energy together because renewable energy is one of the means to reduce greenhouse gases, in this case mostly carbon dioxide (CO_2), which will retard global warming, which is the type of climate change that everyone is worried about. So, renewable energy and climate change go hand-in-hand.

The Beginnings

The last ice age ended about 20,000 years ago. Since then, the Earth has been slowly warming. Furthermore, as the Sun's rays strike the Earth, there are five gases in the atmosphere that absorb the rays as they bounce off the Earth's surface and re-radiate them back into the atmosphere. They are water vapor, CO_2, methane, nitrous oxides, and ozone. They work just like the glass in a greenhouse: they trap heat.

There are many natural processes that release these gases into the atmosphere, but we humans add over 45 billion tons of CO_2 a year to this process. Does this add to the "greenhouse effect"? Yes. Does it make the planet warmer? Yes. Is it a good idea to scale back on these emissions in order to retard the rate of global warming? Yes, it certainly is. Scaling back on carbon emissions is going to cost money. Enter environmental finance.

Climate change, greenhouse effect, greenhouse gases, and *global warming*—these are words that entered our collective consciousness with the "Earth Summit," which was held in Rio de Janeiro in 1992. This meeting is officially known as the United Nations Conference on Environment and Development (UNCED). It is also called the "Rio Summit."

But the genesis of this concern, as well as a more generalized concern for the environment itself, occurred in the late 1960s, both in the United States as well as around the world. In 1969, the United Nations (UN) General Assembly decided to convene a conference to formulate policy alternatives for governments facing environmental issues. And so, the United Nations Conference on the Human Environment (UNCHE) occurred in Stockholm in 1972, the same year that the US Congress passed the Clean Water Act.

Eleven years later—now even more haunted by the impact of industrialization on the environment—the UN Secretary General created the World

Commission on Environment and Development (WCED), which soon took on the name of the Commission's dynamic chair, the former Norwegian Prime Minister, Gro Harlem Brundtland, and became known at the "Brundtland Commission."

In October 1987, the Brundtland Commission issued its report titled "Our Common Future." In this document, they defined—for the first time—the term *sustainable development*. They defined sustainable development as "development that meets the needs of the present without compromising the ability of future generations to meet their own needs."

In addition, "Our Common Future" set the stage for the "Earth Summit" in Rio de Janeiro in 1992, where the common concept of climate change was born.

Two of the major issues taken up in Rio were (1) alternative sources of energy to replace fossil fuels linked to global warming and climate change, and (2) a new reliance on public transportation to reduce vehicle emissions. The Rio Summit, which lasted 6 months, produced two major declarations: (1) "Agenda 21," which was a blueprint/action agenda for achieving sustainable development, and (2) the "Rio Declaration on Environment and Development," consisting of 27 principles to guide sustainable development.

But, by far, the most important outcome of the "Earth Summit" was the negotiation of the United Nations Framework Convention on Climate Change (UNFCCC). The purpose of this groundbreaking treaty was "to stabilize greenhouse gas concentrations in the atmosphere at a level that would prevent dangerous anthropogenic interference with the climate system." As of today, there are 195 nations that are signatories to this agreement.

One of the provisions of the UNFCCC is that the parties would convene each year to assess progress. These meetings are called Conferences of Parties (COPs). In 1997, the COP was held in Kyoto, Japan, where a further agreement was reached, known as the infamous "Kyoto Protocol."

The Kyoto Protocol set binding limits for developed countries on the emission levels of six greenhouse gases: carbon dioxide, methane, nitrous oxide, sulfur hexafluoride, hydrofluorocarbons (HFCs), and perfluorocarbons (PFCs). The emission limits were set at 1990 levels. In other words, each developed country was to reduce the emissions of these six gases to the levels they were in 1990.

The Kyoto Protocol was ratified by all members of the United Nations except Afghanistan, Andorra, Canada, South Sudan, and, most notably, the United States of America.

In addition, the Kyoto Protocol established a system of assigning allowances of emissions, especially CO_2, and permitted the trading of these allowances, or credits. This resulted in what the British call the "European Trading Scheme," and the rest of us anglophones, who are leery of the word *scheme*, call the European Trading System (ETS).

As part of this trading scheme, the Kyoto Protocol also created what is called the "Clean Development Mechanism (CDM)." This is a program where, for example, a developing country wants to build a fossil-fueled power plant and an industrial company in a developed country is faced with emission reduction requirements that would be overwhelmingly expensive to comply with. Let us assume that the cost of reducing emissions from the developing country's power plant is far less expensive than reducing the industrial company's emissions. In this case, the industrial company and the developing country's power plant would enter into an agreement wherein the industrial company would pay for the emission reduction system on the power plant and, in return, would receive the credits for the emission reductions. These credits are called Certified Emission Reduction (CER) units. The Executive Board of the UNFCCC Secretariat must certify all CERs before they can be traded. With these certified CER units in hand, the industrial company could obviate the need for the expensive retrofit on their facility.

The Kyoto Protocol also created a similar mechanism called Joint Implementation (JI). This is a system where an Annex I country can invest in a project in an Annex B country and receive the credits, called Equivalent Reduction Units (ERUs). Annex B countries are referred to as countries with "economies in transition" (EIT). Both Annex I and Annex B countries have emission limits. So, JIs take place where an Annex I country finds it cheaper to invest in pollution reduction in an Annex B country than in its own. The Russian Federation and Ukraine were expected to be major participants in the JI program.

The final chapter, for our purposes, in the history of the global climate change/renewable energy movement occurred at the 15th COP held in Copenhagen in the autumn of 2009. At this meeting, the parties pledged to create an *annual* fund of $100 billion, starting in 2020, to pay for climate change projects. The sad irony of this gesture is that the poor, developing countries are thinking that this $100 billion a year is going to be grant funds. In my humble opinion, it will be a "cold day in hell" before the developed countries marshal $100 billion a year for climate change projects, and an even colder day if they are grant funds.

All of this underscores, however, the need for the wisest environmental finance policies to ensure that whatever funds are provided are done so at the lowest possible cost.

Finance Policy for Climate Change and Renewable Energy

What does all of this have to do with finance policy? How does finance policy fit into this mix of climate change and renewable energy? The answer to these questions is deceptively simple.

You may recall from reading above that two of the most important issues that were taken up at the Rio Summit were alternative sources of energy to replace fossil fuels and a new reliance on public transportation to reduce vehicle emissions.

How do we replace fossil fuels? We have to make fossil fuels more expensive and renewable energy less expensive.

How do we get people out of private vehicles and onto public transportation? We make vehicle ownership more expensive and public transportation less expensive.

As you can see, the answers to these questions *are* simple. It is putting them into practice that is monstrously difficult.

Before saying anything more, we need to note that both renewable energy and public transportation not only have to be less expensive, but they must also be extremely reliable. This leads us into a discussion of renewable energy.

In the past ten years, windmills and solar panels have become the "poster children" of renewable energy. The US Department of Energy created programs of financial incentives for the installation of solar arrays or windmills, or both. The famous PACE (Property Assessed Clean Energy) program was originally designed for solar power but was eventually broadened to include other renewables as well as energy efficiency projects.

The problem with solar and wind is reliability.* When the sun does not shine, no solar power. When the wind does not blow, no wind power. Furthermore, a decade ago there were high hopes that there would be major breakthroughs in the energy storage, or battery, industry. These did not materialize, as many unhappy investors can attest.

That said, solar and wind power may not be the panacea for our fossil fuel problems, but they certainly have a major role to play. In areas like the southwestern United States, where there is abundant sunshine coupled with exceedingly high temperatures, solar power can reduce the cost of air conditioning. Furthermore, in the rural areas of many developing countries, solar and wind can provide electricity where there is none and where settlements are miles from the local grid.

What types of finance programs can provide the greatest incentive for installing solar or wind (or other) renewable energy sources?

The PACE program, which we discussed in Chapter 1, is a great example. *PACE* has a nice ring to it, but as an acronym, it is gawky. It stands, of course, for Property Assessed Clean Energy, which does not really convey a lot of information. Essentially, the PACE program is one where you install solar panels on your roof. You take your bill for these panels to City Hall where you get a check to pay for them. In return, you agree to pay back the city over the next 20 years. The "property assessed" concept comes in because the city slaps a lien on your property to secure their loan to you. The really elegant aspect of this program is that if you sell your house with the solar panels on it, the new owner gets the benefit of the power it produces, *and* the new owner gets to make the annual payments that are part of his/her property tax bill.

* It is also the problem with tidal power.

Contrast this with paying for the solar panels with a second mortgage. Let us say the bank gives you a 7-year second mortgage for $20,000 at 7% interest. As such, your annual payment would be $3,711, or about $309 a month.

With the PACE program, you would get a $20,000 loan for 20 years at about 5%. The annual payment to the city, in this case, would be only $1,605, or about $134 a month.

Which would you rather have? So, you see, PACE-type programs can provide powerful financial incentives to make environmentally smart decisions.

In Chapter 3, we described another similar situation, this time involving a geothermal project in upstate New York. Here, a wealthy gentleman owned several adjoining properties and had a geothermal system installed on all the buildings. The installations cost him $60,000, which he paid in cash. His contractor reported that the gentleman was thrilled to see a $1,000 per month reduction in his electricity bills. She bragged that the gentleman would get all his money back in 60 months, or five years!

How many people do you think are willing to wait five years to get back their money from an environmental investment? Very few.

But, as Chapter 3 explains, if the gentleman's county had a PACE-type program, the man could have taken his $60,000 bill to the county and borrowed the money to pay his contractor. Then the county would have issued a bond to pay for the man's project as well as (hopefully) many others. Geothermal systems last 30 years. So, the county's bond would have issued a 30-year bond at about a 5% interest rate. As such, the gentleman's monthly (re-)payment to the county would have been about $325 a month. In other words, *he would have paid $325 a month, but he would have saved $1,000 a month.* So the man would have had a profit or gain on day one, instead of waiting five years.

Now let us add another layer to these two examples. Let us say that your solar panels save you $150 per month in electricity charges. That is $1,800 a year. And the cost of the solar array is only $1,605. So, you are saving money.

But what if your savings are only $100 a month, or $1,200 a year? What if the New York gentleman's savings were only $200 a month instead of $1,000? In this case, your solar array is costing you an extra $405 a year. Furthermore, you do not get back your money on the geothermal system in 5 years; you never get it back! Does this make you want to install solar panels or geothermal systems? No. What to do? How do we create additional incentives for solar and wind (and other) renewables?

Answer: (ugh!) Subsidies!

You have hopefully read the chapter in this book entitled "The Curse of Subsidies" (Chapter 15), so you know how pernicious, wasteful, and destructive they can be. But you have also hopefully read the chapter on "Grants and Affordability" (Chapter 13). Two of the four *legitimate* uses of grants are to make environmental projects affordable and to introduce new technologies. Both of these principles apply to the use of subsidies (grants) for solar and wind power.

In the case of solar arrays, if a $5,000 grant were available, the cost of the panels could be reduced to $15,000. At that level, the annual payments to the city* would be $1,200, or an amount equal to the savings on the electricity bill.

So, is this justifiable? Yes. It is the use of grant funds to introduce a new and needed technology.

And, yes, it also makes these solar panels affordable. Usually, when we talk about affordability, we talk about the concept in terms of the poor who were unable to meet their payments because of their poverty. But in Chapter 13, we said there were four legitimate uses of grants. One of these was "incentivizing people to undertake valuable environmental projects." So, in this case, we are talking about affordability in the sense of getting people, who are not necessarily poor, to undertake environmental improvements. If they can save money or, at least, break even, then many people will undertake such projects. If they can't, they won't.

So, we must use grants judiciously, in conjunction with other sound environmental finance programs, to incentivize people to make environmental improvements. This is how we are going to advance the cause of renewable energy.

A final note about introducing new renewable energy technologies.

As noted above, the sun does not always shine. The wind does not always blow. The tides only ebb and flow twice a day. Biomass from substances such as switchgrass is only available when it is harvested.

On the other hand, hydroelectric dams produce power 24/7. So do modular nuclear plants, like the ones in naval ships. So does geothermal. So do biofuels from sources such as animal wastes and algae. So do ocean currents (which along with algae may be the real sleepers of future renewables). So do fuel cells. And so does nuclear fusion.†

Our scientists and other experts on these matters need to decide which of these technologies show the most promise for relieving us of the greenhouse gases imbedded in our fossil fuels. Then they must use grants to develop and introduce these technologies.

Now let us turn to the other major issue at the Rio Summit: Getting people out of vehicles and into public transportation.

There are three major forms of urban mass transport: metro (or subways), Light Rail Transport (LRT), and Bus Rapid Transport (BRT). Of these, BRT is, by far, the least expensive.

That said, it is not cheap. So from an environmental finance policy perspective, our primary goal is to achieve the lowest possible Annual Debt Service Payments (ADSPs) on BRT projects in the assurance that the lower the cost, the greater will be government interest in undertaking such projects.

* On a $15,000 loan financed by a municipal bond for 20 years at a rate of 5%.

† In the 1970s, the New York State Science and Technology Foundation made a substantial grant to one of the State's major universities to develop commercial-scale nuclear fusion. We are all still waiting for results.

Achieving the lowest ADSP on projects in developing countries most often depends on lengthening the term of financing, rather than seeking out the lowest possible interest rate. Interest rates play an important role, but not as much as term. Lengthening the term on a $100 loan (at 6% interest) from five to six years decreases the annual payment from $23.74 to $20.34, a decrease of 14%; whereas reducing the interest rate from 6% to 5% only reduces the annual payment from $23.74 to $23.10, a decrease of only 3%.

Most projects in developing countries are either at the mercy of their domestic banking industry or the international development banks (IDBs). One of the principal characteristics of banking systems in developing countries is truncated term. There are several reasons for this. The first is illiquidity. Borrowers' financial accounts in developing countries are rarely maintained in accord with International Financial Reporting Standards (IFRS). This necessarily results in opaque loan documentation, unacceptable to secondary markets. Such loans are, therefore, illiquid. Once a bank makes a loan in a developing country, it is generally stuck with that loan through its full term.

For many of the same reasons, projects in developing countries have little or no access to the international capital markets. In addition to nonstandard accounting and opaque loan documentation, the rule of law and the enforceability of contracts are also an issue. Finally, currency exchange is another problem. It is one thing to provide a spot market for thin and unfamiliar currencies; it is another thing, entirely, to do so for long terms.

There is an exception to this general rule: Portuguese banks have a strong appetite for Brazilian debt, and Spanish banks have an appetite for project finance throughout Latin America. Over the past 20 years, Spanish and Portuguese banks have sought superior returns by financing projects in emerging markets where mainstream international banks have feared to tread. In doing so, they have been burned on several occasions, most notably in the republics of the former Soviet Union. However, their appetite for Brazilian and Latin American debt remains strong.

To reach the international credit markets, and thus obtain the longest terms and lowest interest rates, urban mass transport projects must use *credit enhancement*. These should include Self-Funded Reserves and Second Loss Reserves. It should also specifically involve a new role for the IDBs: purchasing the first loss tranches of these types of financings. (All these concepts are more fully set out in Chapter 18 on "Credit Enhancement.") In addition, projects such as these in developing countries should explore using either guaranties from Export Credit Agencies (ECAs) (to guaranty payments to foreign vendors of equipment), the Multilateral Investment Guaranty Agency (MIGA), or private political risk insurance. A solid guaranty and/or political risk insurance will be the additional support that is needed to attract investment from the international capital market.

Finally, any equity, either as a result of a public–private partnership or otherwise, should be avoided if possible. The first and foremost reason is the cost of equity. New venture capital commands a compound rate of return in the 20% to 25% range. For developing countries, this range can reach as high as 40%. This is inconsistent with the goal of providing the lowest cost of financing. It should only be considered on the rare occasion when a private business might be an essential element of a project, and having that company's skin in the game is essential to obtaining its cooperation.

In many BRT projects, the bus fares will cover the cost of the buses, but not the infrastructure. In such cases, subsidies from the government are needed to operate the system and finance the infrastructure. The key to obtaining the best terms and conditions on the debt is to tie together the financings of the buses and the infrastructure with cross-default clauses in all project documents. This means that there will be two streams of payment: bus fares and subsidies. This is crucial. The local transit authority will then pledge both revenue streams to the repayment of the project debt. Most importantly, however, the subsidy-granting unit of government will be asked, in a three-party agreement, to pay the subsidies directly to the project fiscal agent in the event the local transit authority defaults on a payment. The failure of government to do so is a political act, which can be covered by political risk insurance. The relationship between the lender/investor and the local transit authority is a debtor/creditor relationship involving pure commercial risk. But the relationship between the higher unit of government and the investors does not involve commercial credit risk. Instead, it involves political risk.

So, the international capital markets offer the longest terms and the lowest interest rates. The key to reaching these markets is to wrap a project financing with credit enhancements including political risk insurance from either MIGA or a private insurer.

One of the major advantages of going to the international capital markets is the ability to finance projects over the service lives of their components.

In general, BRT projects can be divided into four major components based on the concept of service life. Service life governs the maximum term for which an asset can be financed. The four components are (1) buses, (2) infrastructure and depot buildings, (3) heavy equipment, and (4) a category that includes vehicles, office equipment, and furnishings. The corresponding service lives for these categories of assets are as follows:

1. Infrastructure and depots: 20 years
2. Buses: 12 years
3. Heavy equipment: 10 years
4. Vehicles, office equipment, and light equipment: 5 years

Below is a table that illustrates the annual payments (at a 10% interest rate) that would be due on a model project with a 20-kilometer route, which would require 150 buses at $300,000 ($45,000,000 total) each to serve 150,000 to 200,000 passengers per day. Such a project would require about $55,000,000 of roads and buildings, about $5 million of heavy equipment, and $5 million of light equipment for a total project size of $110,000,000.

	5 years	10 years	12 years	20 years
Infrastructure (20 yrs)	$14,508,861	$8,950,997	$8,071,982	$6,460,279
Buses (12 yrs)	11,870,887	7,323,543	6,604,349	6,604,349
Heavy equipment (10 yrs)	1,318,987	813,727	813,727	813,727
Light equipment (5 yrs)	1,318,987	1,318,987	1,318,987	1,318,987
Total	**$29,017,722**	**$18,407,254**	**$16,809,045**	**$15,197,342**

It is clear from the above table that lengthening the term to coincide with the service life of the underlying asset class has the greatest positive effect on the clients' required cash flow.

Now the final question for our consideration in this chapter is: Where will the money come from for all the grants and subsidies? The answer is: from broad-based income streams. This is never going to be popular but, sooner or later, it has to happen. As of this writing, a visiting alien from another planet would conclude that we must have a climate change policy. If we did, in the area of renewables, we would subsidize the new technologies by taxing the nonrenewables: *a carbon tax.*

In like manner, if you want to get people out of cars and onto public transportation, you should make driving a car more expensive and taking public transit less expensive. There are many ways to make driving a car more expensive. Some of them were discussed earlier in this book. One is certainly a vehicle emissions tax where the more pollution a vehicle emits, the more its owner has to pay in the form of a tax. The other way to make driving more expensive is to increase the tax on gasoline and use the proceeds to reduce the fares on public transportation.

There are also ways to attack other specific climate change problems. Take the case of carbon and other emissions from the dray trucks, which were discussed in Chapter 1, that plague every port in the world. To address this issue, every container or ton of freight that enters a port could be taxed. The proceeds could then be used (1) to replace the old dray trucks, (2) to clean up the water pollution and stormwater problems at ports, and (3) for dredging and other port improvement projects. And, in order to keep the North American continent on a level playing field, thereby preventing shippers from choosing the ports of one country over another, such a tax could be implemented by amending the North American Free Trade Agreement (NAFTA) so that the tax would be collected in Canada and Mexico as well as the United States.

At the international level, there are fewer options. This is because there are two sources of opposition: public unpopularity and intergovernmental mistrust. Some world governments clearly do not trust other world governments, especially when it comes to handling money.

Perhaps the only palatable international tax would be one on airline freight and airline passenger miles. As noted in Chapter 7, in the 2010 flight year (May 2010 to April 2011), there were 817,743,551,000 air passenger miles flown in the United States alone, including both domestic and international flights*. If passengers were charged only 1 cent for each mile they flew, that would generate more than $8 billion each year—just in the United States. Ten cents a mile would probably not terribly depress air travel. That would bring in $80 billion in the United States alone. If those funds were used as 100% subsidies, they could pay for over $1.2 *trillion* of climate change or renewable energy projects.

In addition, in 2010, there were 7,304,348,000 ton/miles of freight flown just in the United States. At, $1 per ton/mile, a tax would generate another $7.3+ billion, enough to finance well over $100 billion of renewable energy or climate change projects.

In conclusion, we are going to have to reduce our man-made emissions of greenhouse gases. A major part of the strategy for doing so will be to replace fossil fuels with nonpolluting renewables. Smart subsidies will be essential for making these changes in the most efficient and painless ways possible.

What will not be painless is raising the money for the needed subsidies. However, we will make no meaningful progress in retarding global warming until we drum up the courage to do so.

* US Bureau of Transportation Statistics, US Air Carrier Traffic Statistics through April 2011, available at: http://www.bts.gov/xml/air_traffic/src/index.xml.

21

Cap-and-Trade Programs

Cap-and- trade programs are attempts to let free-market forces lower the cost of reducing targeted pollutants. From a finance policy point of view, there is not much to say or do about cap-and-trade programs, other than to stand on the sidelines and cheer. When they work, they lower the cost of reducing emissions. The problem, as you will see, is that they do not always work.

In environmental finance, there are two largely complementary approaches: (1) traditional project finance and (2) market mechanisms. With limited exceptions, capital improvement projects for all environmental media are financed through traditional project finance methodologies—raising debt and equity—with the important twist that lowest cost, not maximum return, is the goal.

In the field of air quality, and more recently in water quality, there is an additional mechanism: the free market. This market mechanism is based on *cap-and-trade programs*. The total amount of pollutants that may be emitted by eligible participants is "capped" by government at a level sufficient to achieve specific policy goals. *Allowances, offsets,* or *credits* are then created and either purchased from the government or allocated by the government to polluters.

So, Polluter A might emit 100 tons of, say, sulfur dioxide (SO_2), but get an allowance from the government of 90 tons. Polluters are required to undertake reductions in emissions and, to the extent they reduce emissions below their limit, they can sell their excess allowances. So, Polluter A can undertake a project to reduce its emissions. If they reduce them to, say, 80 tons, they can sell the difference between the 90 tons they are allowed and the 80 tons they actually emit. Thus, they can sell 10 tons of allowances. Conversely, another polluter, Polluter B, facing a more expensive pollution reduction project, might find it less expensive to purchase Polluter A's allowances rather than undertake a more costly project on its own.

There are two financial effects of this mechanism. First, the proceeds of the allowance sales reduce the cost of the seller's (Polluter A's) pollution reduction project. And, second, it reduces the buyer's (Polluter B's) cost of compliance. So, the total cost of pollution reduction with Polluters A and B trading credits is lower than the cost of each undertaking its own project. Recall that the name of the game is driving down the cost of environmental improvement projects.

A similar process is evolving in the clean water space, specifically in the Chesapeake Bay watershed, which is one of the nation's most sensitive water bodies. In the near future, we will begin to see communities that want to grow, but cannot because of the permitted capacity of their water treatment plants. Each wastewater treatment plant in the country operates under a National Pollutant Discharge Elimination System (NPDES) permit. When the plant reaches the limit specified in its permit, it cannot accept more sewage. No new connections. No growth. In such case, a developer wanting to add, say, 500 homes will have to find and buy "credits" equal to the capacity he is adding to the system. He will then turn over these credits to the wastewater treatment serving his development. The wastewater treatment plant will only be able to operate over its permit limit if it shows the regulators that it has credits equal to the overage. There are several programs throughout the country where point-source polluters, such as sewage treatment plants, can trade credits. But the only state so far that has undertaken a cap-and-trade program that deals with both point-source as well as non-point-source pollution is the Commonwealth of Pennsylvania. This program was organized and is managed by the Pennsylvania Infrastructure Investment Authority (PennVest).

PennVest has done an excellent job and has had several successful auctions of credits, but its program operates under some state-imposed constraints that speak to larger problems with cap-and-trade programs. Pennsylvania is drained by two massive water systems. In the east, it is the Susquehanna River; in the west, it is the Allegany and Monongahela Rivers that come together in Pittsburgh to form the Ohio River. The Commonwealth of Pennsylvania will not permit credit trades between the two watersheds. Furthermore, the Susquehanna is a truly massive watershed in Pennsylvania, but there is a little sliver of the south-central part of the state that is in the Potomac River watershed. Now, both the Susquehanna and the Potomac Rivers are part of the Chesapeake Bay watershed, but the Commonwealth will not permit trades between these two systems within the same watershed.

So, you can see some policy workings going on here. The Commonwealth does not want to have some dirty rivers where permit compliance is very expensive (and, hence, credits are purchased instead of clean-ups) and some clean rivers where compliance is cheap.

This speaks to a problem that has vexed other cap-and-trade programs other than just PennVest's nutrient trading program. The problem is *hotspots*.

In 2005, the US Environmental Protection Agency (EPA) issued its Clean Air Interstate Rule (CAIR), which functionally set up a cap-and-trade program for SO_2 and nitrogen oxides (NOx). This program got off to an ambitious start. Credits were actively traded on the Chicago Mercantile Exchange (CME).

Now, both SO_2 and NOx are *criteria pollutants*, which means they can impair health and welfare. Nonetheless, the trading program continued apace until one day some polluter bought some SO_2 credits instead of installing more expensive abatement equipment. Then, someone living downwind of this polluter sued the polluter because he had to breathe the lousy air instead of better air if the polluter had installed abatement equipment. The courts sided with the plaintiff and threw out the CAIR.

The EPA then came back with what might be called "son of CAIR," titled the Cross State Air Pollution Rule (CSAPR), which is dyslexically pronounced "casper." The courts have also voided the CSAPR because of the same problem: The downwind area is a "hotspot." The courts will not permit hotspots. As of this writing, the CSAPR suit is on petition to the US Supreme Court.

I am not a prophet—and certainly not when it comes to cases in court. However, I do not see how the CSAPR will survive. Neither do I think nutrient trading programs like PennVest's will survive. As much good as these cap-and-trade programs do in decreasing the cost of reducing pollutants, I just cannot see the courts allowing you to foul my downwind air or downstream water just because you bought a cheap credit from someone. And, that's about it.

Carbon trading, however, is another matter. CO_2 does not produce hotspots. Carbon is a global problem. So, there will be no plaintiffs complaining of upwind emitters. And there will be no courts issuing injunctions against carbon credit trading programs.

That said, carbon cap-and-trade programs have their own problems. The two major problems with carbon credit (allowance) trading deal with maintaining the price in the face of two uncertainties: incorrect estimates and economic fluctuations.

Several of the carbon trading programs have foundered because of overestimating the "cap," thereby making the credit worth much less because, with a high cap, there were fewer entities that needed credits. The Regional Greenhouse Gas Initiative (RGGI), a program of nine mid-Atlantic states in the United States, foundered for this reason. In 2013, the RGGI reorganized with tighter caps and floor prices.

The second problem with carbon trading programs is economic fluctuations. In good economic times, industry is booming and everyone needs credits; in bad times, it is the opposite: No one needs credits. In the former case, carbon prices spike up; in the latter case, they collapse. That was part of the RGGI's problem too, but its most important victim was the European Trading Scheme (ETS), which collapsed in January of 2013. As of this writing, the resuscitation of the ETS is in the works, and the California Air Resources Board (CARB) has opened an ambitious new carbon-trading program. CARB spent a lot of time and effort learning from the mistakes of the other trading programs. CARB is one of the best and most sophisticated air agencies in the world. Odds are that they will succeed.

Accurately counting tons of carbon is a finite and doable job. Sooner or later, they will get it right. But dealing with economic ups and downs is another matter. However, with enough flexibility, and with strategies like floor pricing where the agency buys the credits at a known price when demand is slack, they will ultimately succeed.

Note that there is an important lesson to be learned from the principle of providing the greatest environmental good for the largest number of people at the lowest possible cost. That lesson is: The lower the cost of creating an environmental benefit, the more environmental benefits will be created. If the cost of purchasing credits, or allowances, is low, more environmental benefits will ensue. In parallel, if the cost of undertaking an environmental improvement project is low, more such projects will be undertaken.

As such, to accomplish environmental quality improvement goals, environmental finance must (1) in the case of cap-and-trade programs, increase the cost of allowances; and (2) in more traditional project finance, reduce the cost of environmental improvement projects.

22

Driving Down Costs

This book began by saying that the overarching goal of environmental finance is to provide the greatest environmental benefit for the largest number of people at the lowest possible cost. That is Core Principle 1. Core Principle 2 is driving down the cost of environmental improvement projects.

In the preceding pages you have read, sometimes perhaps *ad nauseam*, about how you actually drive down costs—how you get from A to B.

So, what I would like to do now is to leave you with a short summary of the more important factors to consider when formulating environmental policy, especially when confronting the question of how to pay for environmental improvements.

It goes without saying that driving down costs should be accompanied by efforts to increase Cash Available for Debt Service (CADS). This should be a major element of any environmental finance policy.

Reduce System Costs

The starting point for increasing CADS is reducing system costs. When the US Environmental Protection Agency (EPA) began studying its $70+-billion construction grant program that ran from 1972 to 1987, they found a lot of waste. There was overbuilding on the project side and many wasteful practices on the operational side. Why not? The massive grants were free money. No one was going to notice some fat in the operating budget. Do not let that happen. Get some comparisons. Bring in some experts to evaluate system operations.

Volumetric Tariffs

Make sure environmental utility systems base their rates on increasing block tariffs (IBTs). The next gallon of water must be more expensive than the previous one. This will increase CADS and, to some extent, encourage conservation as well.

Raise Money/Change Behavior

As noted above, IBTs change behavior (that is, encourage conservation) while they are raising money. So should other policies. Emission taxes on automobiles will drive consumers to more efficient, less polluting vehicles. So will increased gasoline taxes. So will carbon taxes.

Service Lives

Always finance new facilities in accordance with their service lives. People who enjoy benefits in the future should pay their fare share for them.

Longest Terms

When seeking financing alternatives, look for the ones with the longest terms. They are the ones that will result in the lowest annual payments for your ratepayers.

Lowest Rates

Remember that interest rates are a function of risk and time. Long-term rates are higher than short-term rates. But using debt with longer terms reduces annual payments far more than the modest increases in interest rates over time. So, it is not the time, but rather the risk end of the rate game that you must look to. Drive down interest rates by driving down the risk of loss.

Credit Enhancement

Chapter 18 describes several strategies for reducing interest rates with credit enhancement mechanisms. As you are preparing to finance a project, or vote on the structure of an environmental finance program, you should examine each strategy and use it if it can help lower your interest rates.

Guaranties

Guaranties are probably the single most important factor in reaching the greatest number of people with low-cost environmental benefits. Remember the numbers from the simulation in Chapter 16: You can only make 81 subsidized loans with the same amount of money that can guaranty 543 projects.

Buy-Downs

When dealing with projects that have affordability issues, use grants to buy down the cost of the project to affordable levels.

Subsidies

Use subsidies only for the truly needy and to induce people to undertake nonmandatory environmental improvements. Try to get rid of the general, supply-based subsidies for the rich and un-needy.

Carrots and Sticks

When you are preparing to regulate, have an eye to creating incentives for compliance with your regulations—especially for the needy.

Cost/Benefit

The name of the game, as you well know by now, is to *bring the greatest amount of environmental benefits to the largest number of people at the lowest possible cost.* Using cost/benefit analyses on all the projects in your program will enable you to identify precisely which project is the one that will *bring the greatest amount of environmental benefits to the largest number of people at the lowest possible cost.*

* * * * * *

I hope that the contents of this book will convince you of the need for people to pay attention to finance when formulating environmental policy. As has been said many times in this book, *the goal of environmental finance policy should be to bring the greatest amount of benefits to the largest number of people at the lowest possible cost.* And the means for doing so is to use some of the strategies discussed here to drive down the cost of environmental improvement projects.

Appendix: Countries with Non-Investment-Grade Ratings on Their Sovereign Debt

Below is the list of countries that, according to Standard & Poor's, have non-investment-grade (below BBB–), or "speculative," credit ratings, and their respective ratings.

Albania	B+
Angola	BB–
Argentina	B–
Bangladesh	BB–
Barbados	BB+
Belarus	B–
Belize	B–
Benin	B
Bolivia	BB–
Bosnia	B
Burkina Faso	B
Cambodia	B
Cameroon	B
Cape Verde	B+
Cook Islands	B+
Croatia	BB+
Dominican Republic	B+
Ecuador	B
Egypt	CCC+
El Salvador	BB–
Fiji	B
Gabon	BB–
Georgia	BB–
Ghana	B
Greece	B–
Guatemala	BB

Honduras	B+
Hungary	BB
Indonesia	BB+
Jamaica	CCC+
Jordan	BB
Kenya	B+
Lebanon	B
Macedonia	BB
Mongolia	BB–
Montenegro	BB–
Mozambique	B+
Nigeria	BB–
Pakistan	B–
Papua New Guinea	B+
Paraguay	BB–
Portugal	BB
Romania	BB+
Rwanda	B
Senegal	B+
Serbia	BB–
Sri Lanka	B+
Suriname	BB–
Tunisia	BB–
Turkey	BB+
Uganda	B+
Ukraine	BB
Venezuela	B
Vietnam	BB–
Zambia	B+

Index